CONJUGATED POLYMERS
AND OLIGOMERS
Structural and Soft Matter Aspects

Energy and sustainability are keywords driving current science and technology. Concerns about the environment and the supply of fossil fuel have driven researchers to explore technological solutions seeking alternative means of energy supply and storage. New materials and material structures are at the very core of this research endeavor. The search for cleaner, cheaper, smaller and more efficient energy technologies is intimately connected to the discovery and the development of new materials.

This collection focuses on materials-based solutions to the energy problem through a series of case studies illustrating advances in energy-related materials research. The research studies employ creativity, discovery, rationale design and improvement of the physical and chemical properties of materials leading to new paradigms for competitive energy-production. The challenge tests both our fundamental understanding of material and our ability to manipulate and reconfigure materials into practical and useful configurations. Invariably these materials issues arise at the nano-scale!

For electricity generation, dramatic breakthroughs are taking place in the fields of solar cells and fuel cells, the former giving rise to entirely new classes of semiconductors; the latter testing our knowledge of the behavior of ionic transport through a solid medium. In energy-storage exciting developments are emerging from the fields of rechargeable batteries and hydrogen storage. On the horizon are break-throughs in thermoelectrics, high temperature superconductivity, and power generation. Still to emerge are the harnessing of systems that mimic nature, ranging from fusion, as in the sun, to photosynthesis, nature's photovoltaic. All of these approaches represent a body of materials–based research employing the most sophisticated experimental and theoretical techniques dedicated to a common goal. The aim of this series is to capture these advances, through a collection of volumes authored by leading physicists, chemists, biologists and engineers that represent the forefront of energy-related materials research.

For further details, please visit: http://www.worldscientific.com/series/mae

(Continued at the end of the book)

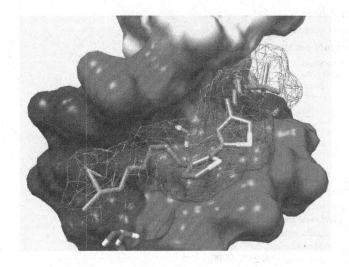

MATERIALS AND ENERGY – Vol.9

CONJUGATED POLYMERS AND OLIGOMERS

Structural and Soft Matter Aspects

Editor

Matti Knaapila

Technical University of Denmark

 **World Scientific**

NEW JERSEY · LONDON · SINGAPORE · BEIJING · SHANGHAI · HONG KONG · TAIPEI · CHENNAI · TOKYO

Published by

World Scientific Publishing Co. Pte. Ltd.

5 Toh Tuck Link, Singapore 596224

USA office: 27 Warren Street, Suite 401-402, Hackensack, NJ 07601

UK office: 57 Shelton Street, Covent Garden, London WC2H 9HE

Library of Congress Cataloging-in-Publication Data

Names: Knaapila, Matti, editor.

Title: Conjugated polymers and oligomers : structural and soft matter aspects / editor,
   Matti Knaapila (Technical University of Denmark, Denmark).

Other titles: World Scientific series in materials and energy ; v. 9.

Description: Singapore ; Hackensack, NJ : World Scientific, [2017] |
   Series: Materials and energy, ISSN 2335-6596 ; vol. 9

Identifiers: LCCN 2017030402| ISBN 9789813225756 (hardcover ; alk. paper) |
   ISBN 9813225750 (hardcover ; alk. paper)

Subjects: LCSH: Conjugated polymers. | Oligomers.

Classification: LCC QD382.C66 C696 2017 | DDC 547/.70457--dc23

LC record available at https://lccn.loc.gov/2017030402

British Library Cataloguing-in-Publication Data

A catalogue record for this book is available from the British Library.

For any available supplementary material, please visit
http://www.worldscientific.com/worldscibooks/10.1142/10591#t=suppl

Desk Editor: Rhaimie Wahap

Typeset by Stallion Press
Email: enquiries@stallionpress.com

Printed in Singapore

Preface

Soft materials are characterized by self-organized structures at mesoscopic length scales well above the atomic level. Structural transformations take place at energy scales comparable to room temperature thermal energies. Conjugated polymers and oligomers have found widespread applications from OLEDs to sensors. In addition to photophysics, charge transport and device fabrication, this field has also been paralleled by phase behavioral studies since the first structural reports of polythiophenes in the late 1980s by Winokur, Tashiro and others which follows from even earlier work. This book identifies current trends of structural aspects of conjugated polymers and oligomers. All chapters are written by leading scientists in their fields, ensuring state-of-art coverage.

Most properties of conjugated molecules and supramolecules stem from their stiff conjugated backbone. The understanding of this fundamental structural unit has become exceedingly advanced. This is described by Wunderlich, Müllen and Fytas in Chapter 1. The area has benefitted greatly from advances in instrumentation such as high resolution transmission electron microscopy and electron diffraction and tomography. This is beautifully discussed by Brinkmann in Chapter 2. Processing conjugated molecules from solutions has made significant progress and now includes water solutions and dispersions. Strategies to control their solubility in water include incorporation of hydrophilic terminal groups into side chains off the polymer backbone as well as surfactants that may form a layer between the polymer and water. This topic is reviewed by Burrows, Stewart, Ramos and Justino in Chapter 3. This has led to an increasingly active trend is to use them in detecting biomolecules and DNA. This requires careful understanding of intermolecular interactions and the subsequent formation of supramolecules. Interactions between conjugated molecules and DNA are discussed by Alemán and Zanuy in Chapter 4. A comprehensive overview of the supramolecular assemblies of conjugated polymers and DNA is provided by Knoops, Rubio-Magnieto, Richeter, Clément and Surin in Chapter 5. The device performance depends on the microstructure and morphology of conjugated materials and these on the other hand depend on the device environment itself. This has led scientists to follow the device manufacturing *in situ* by fitting device processing on synchrotron beamlines. This topic is featured by Andreasen in Chapter 6. Current trends also embrace fundamental studies not directly motivated by device prospects. These niches include for example structural behavior of conjugated molecules under high pressure conditions, discussed by Knaapila, Torkkeli, Scherf and Guha in Chapter 7.

This way each chapter recognizes an active research line where the soft matter perspective dominates the research of conjugated materials. The thesis of the whole book is to introduce a spectrum of themes from fundamental aspects of persistent backbone to water soluble conjugated molecules with surfactants, biomolecules and DNA — and from the advanced use of synchrotron radiation and electron microscopy to the extreme conditions research. I feel that this compact overview will be helpful and inspire and motivate scientists and graduate students in their future work.

I thank the authors for their efforts and the referees for carefully reading the chapters and giving comments that significantly improved the book. I thank Prof. Michael J Winokur of the University of Wisconsin-Madison and Prof. Jean-Luc Brédas of KAUST for the original idea and encouragement in the editorial work.

Matti Knaapila
Department of Physics
Technical University of Denmark
Kgs. Lyngby, Denmark

Contents

Chapter 1

Shape Persistence in Polymers and Supramolecular Assemblies

Katrin Wunderlich,* Klaus Müllen,*,‡ George Fytas *,†,§

*Max Planck Institute for Polymer Research,
Ackermannweg 10, 55128 Mainz, Germany

†Department of Materials Science,
University of Crete and IESL/FORTH,
71110 Heraklion, Greece
‡muellen@mpip-mainz.mpg.de
§fytas@mpip-mainz.mpg.de

A thorough insight into the shape of polymers and supramolecular assemblies is of crucial importance for the flow behavior, the packing in the solid as well as the structure and dynamics in non-dilute solutions. In this chapter, we present the main experimental techniques that can reliably address persistence and compare different polymers with reported persistence lengths ranging from few nanometers to few hundreds nanometers. Specifically, we discuss not only the so-called rigid-rod polymers such as polyphenylenes, but also dendronized and helical polymers. These chain-type macromolecules are generally considered as rigid or semi-rigid compared with conventional homopolymers such as polyolefins. Next it is meaningful to compare these covalently bonded polymers with supramolecular assemblies consisting of amphiphilic polymers such as poly(ethylene glycol) functionalized benzene derivatives and poly(ethylene oxide-*b*-butadiene). Supramolecular assemblies can assume larger rigidity than covalently bonded polymers. Although we can classify different types of polymers and supramolecular assemblies and determine their persistence lengths, so far, there is no clear structure-rigidity relationship. Hence, a prediction of the persistence lengths solely based on the molecular structures remains uncertain.

1. Introduction

The rapidly growing interest in rigid polymers and supramolecular assemblies is due to the chemical challenges and their potential applications in nanotechnology[1] and biomedicine.[2] Toward this end, a detailed analysis of the polymer shape persistence[3] and its impact on the physical properties is of crucial importance. The shape and size of a polymer sensitively depend not only on the monomeric repeating unit, their mode of connection, but also upon weak intermolecular forces (hydrophobic, electrostatic and π-π interactions, hydrogen-bonding, and metal-to-ligand coordination). The persistence length of a polymer chain, a common measure of the flexibility or stiffness of a polymer, is the characteristic length over which the correlation of the tangent vectors along the chain contour drops by about 70%. It can be experimentally determined in dilute solutions by scattering techniques. However, the tendency of persistent polymers towards aggregation, even in the dilute concentration regime, obscures the contribution of individual chains to the time average scattering intensity. The desired resolution is feasible by the dynamic techniques of dynamic light scattering (DLS) and fluorescence correlation spectroscopy (FCS).

The ratio of the contour length L and the persistence length l_p determines the size and shape of the polymer chains. For rigid rods, $L/l_p \leq 1$, whereas for flexible polymers, $L/l_p > 100$. Note that the same polymer chain can fall into both regimes depending on its length, i.e., the molecular weight. While there is a consensus for the two borderline cases, the notion of semi-rigid or semi-flexible chains seems to be vague. To avoid confusion, we deliberately term chains with $L/l_p < 10$ as semi-rigid and as semi-flexible, when $L/l_p > 10$. We also note that the term chain is used both for covalent polymers and supramolecular assemblies. In semi-rigid polymers the spatial correlation of the tangent vectors is long with the persistence length being lower than their contour length. Their structure and dynamics can, in principle, be described by two processes: an out-of-plane bending deformation, which determines the size (end-to-end distance) of the chains, and conformational changes via axial rotations (different dihedral angles, Fig. 1). The impact of the latter on the average chain size depends upon the repeating unit and their connection. The persistence length imparts a wide range of polymer characteristics including self-assembly in dilute solutions, solid-state packing and mechanical properties. Thus, while persistence length of polymer chains is initially determined in dilute solutions and possible aggregation regarded as a "complication", intermolecular forces can become increasingly important

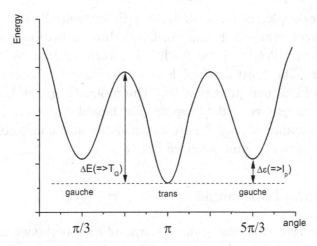

Figure 1: A schematic conformational energy diagram of a polyethylene-like chain.[9] The value of the activation energy (ΔE) controls the mobility associated with the glass transition T_g, whereas the energy difference between two conformations affects the static flexibility parametrized by the persistence length (l_p).

for shape persistence. It is thus appropriate to also consider the transition from the solution to the solid state and to compare covalent polymers with supramolecular assemblies. Such studies are still missing, but based on the distinct flow behavior between flexible and semi-rigid polymers (see Section IV below) are considered necessary.

The objective of this chapter is to compare different, but representative polymers and supramolecular assemblies of different degree of rigidity and to reveal the correlation between structure and their persistence length. First, we present, both, the theoretical background and the experimental techniques: photon correlation spectroscopy (PCS) and FCS. Then we give recent examples of polymers and supramolecular assemblies and compare their persistence lengths. We show that the rigidity of covalent persistent polymers originates either from a rigid aromatic backbone (polyphenylene, polyfluorene), from grafting of side chains of various architectures onto the main chain (linear or branched) or from winding of chains into helices (polyisocyanate, poly(γ-benzyl-L-glutamate)). Complex rigid polymers with linear or branched side chains will be compared with standard conjugated polymers such as poly(para-phenylene) (PPP)[4] or poly(para-phenylene vinylene) (PPV).[5]

To demonstrate that supramolecular assemblies assume distinctly larger rigidity than covalently bonded polymers, we chose from the few reported systems three amphiphiles consisting of a hydrophobic part (either alkenyl or benzene) and a hydrophilic part (amide bonds and/or poly(ethylene glycol)).

In the first example, benzene derivatives functionalized with amides and triethylene glycol form one-dimensional fibrils through hydrogen-bonding and π-π interactions.[6] Poly(ethylene oxide-*b*-butadiene) self-assembles to cylindrical micelles. The third amphiphile is a hexaphenylbenzene-poly(ethylene glycol) (HPB-PEG) derivative (Fig. 9).[7] This molecule was studied because of the assembly propensity of the propeller-like hexaphenylbenzene core under formation of bundles of fibers.[8] Further details about the molecular structure and their persistence lengths are given below.

2. Theoretical Background

Polymers that have rotational symmetry around their backbone are characterized by isotropic rigidity with a single persistence length, l_p, describing the space correlation in the rotation angles, i.e., $\langle t(s)t(0)\rangle = exp(-s/l_p)$, where $t(s)$ is the tangent at a distance s along the chain contour. In a simple rotational angle (trans/gauche) model, there are two important energies, ΔE and, $\delta\varepsilon$, (Fig. 1)[9] that have an impact on the static and dynamic chain flexibility, respectively. For the latter, expressed in the glass transition temperature, is too oversimplified, since it ignores many chain interactions and packing effects. It is nevertheless valid as it decouples the kinetic and equilibrium chain properties. An increase of ΔE is behind all chemical attempts to boost persistence in synthetic polymers. Different from covalent polymers, supramolecular assemblies can possess high rigidity, which is the consequence of enhanced non-covalent interactions and contribution from the out-of-plane bending deformation. We do not address in this review the case of ladder polymers with anisotropic rigidity (in-, and out-of-plane).[10–12] For flexible Kuhn-chains described by the isomeric-state theory,[13] the Kuhn segment length, $l_k = 2l_p$ is determined by $\delta\varepsilon$, whereas a semi-rigid or semi-flexible chains, also known as worm-like Porod chain a bending modulus can be associated[14] to l_p. No solvent effect is considered.

The estimation of l_p is not an easy task since it requires experiments in very dilute solutions, preferably in theta solvents to avoid excluded-volume interactions, and a wide range of scattering wave vectors, q, to access to the full form factor, $P(q)$. Light scattering has lower space resolution than neutron or X-ray scattering but requires no labeling, is sensitive enough to be applied at very low dilution and also provides access to transport coefficients (translational and, less often, to rotational diffusion). While a fully decayed $P(q)$ cannot be obtained by a single scattering technique, the radius of gyration, R_g, can be estimated from light scattering experiments at sufficiently large degree of polymerization ($R_g > 20$ nm). The center-of-mass translation diffusion D_0, however, can always be measured by DLS yielding the hydrodynamic radius, R_h. For optically anisotropic structures, depolarized dynamic light scattering can yield the rotational diffusion, D_R which like $P(q)$ and D_0 are sensitive measures of the conformation. For extremely dilute solutions (in nM concentrations), the fluorescence correlation technique can be employed for measuring D_0 of labeled chains. Imaging techniques, e.g., atomic force microscopy (AFM), transmission electron microscopy (TEM) utilize samples on solid substrates that can often lead to erroneous information about chain rigidity.

A polymer chain with weight-average molecular weight, M_w, monomer length b and monomer molecular weight, M_0, has a contour length of $L = bM_w/M_0$. The mean-square-radius of gyration of monodisperse semi-flexible chains is a function of the persistence length l_p and the contour length L, as given by,

$$R_g^2(L) = L \cdot l_p - l_p^2 + \frac{2}{L}l_p^3 - \frac{2}{L^2}\left(1 - e^{-L/l_p}\right)l_p^4 \tag{1}$$

If L and R_g are known, Eq. (1) allows the estimation of l_p.

An estimation of the persistence length can also be obtained from the analysis of D_0 using relevant models for the diffusion of semi-flexible chains. We first consider the Yamakawa-Fujii (YF) theory of wormlike chain[15,16] that requires the use of a monomer thickness h. Since YF is valid for $L/h > 50$, it can only be applied for longer chains. Alternatively, one can employ a more general expression of D_0 for semi-flexible chains[17] covering larger L/l_p range:

$$D_0 = \frac{k_B T}{3\pi\eta L_w} * \left(1 + \frac{\sqrt{6}}{\sqrt{\pi}L_w}\int_d^{l_p} \frac{L_w - s}{s}exp\left(\frac{-3d^2}{2s^2}\right)ds\right.$$

$$\left. + \int_{l_p}^{L_w} \frac{L_w - s}{\sqrt{2sl_p}}exp\left(\frac{-3d^2}{4sl_p}\right)ds\right) \tag{2}$$

where d is the hydrodynamic diameter of the chain (equivalent to the thickness of YF model). Note that Eq. (2) applies for $l_p/d > 1$.

Equations (1) and (2) are valid for monodisperse chains. To account for finite polydispersity, M_w/M_n, effects of semi-flexible chains the common Schulz-Zimm distribution, w_L,

$$w_L = \frac{L^m}{\Gamma(m+1)}\left(\frac{m+1}{L_w}\right)^{m+1}exp\left[-\frac{L(m+1)}{L_w}\right] \tag{3}$$

with $m = (M_w/M_n - 1)^{-1}$ and $\Gamma(m+1)$ being the gamma function is used in Eqs. (4) and (5) describe the radius of gyration and translational diffusion, respectively:

$$\langle R_g \rangle_z^2 = \frac{L}{12 \cdot l_p} \cdot \frac{m+2}{m+1} - \frac{1}{16 \cdot l_p^2} + \frac{1}{16 \cdot l_p^3 \cdot L}$$

$$+ \frac{(m+1)^2}{L^2} - \frac{(m+1)^{(m+2)}}{L^{(m+2)} \cdot \left(\frac{m+1}{L} + 4 \cdot l_k\right)^m} \tag{4}$$

$$D_{polydisp.} = \frac{\int_0^\infty w_L L D dL}{\int_0^\infty w_L L dL} \tag{5}$$

In case of polydisperse semi-flexible chains, Eq. (1) with $L = L_w$ is known to over-estimate R_g[17]; a polydispersity $M_w/M_n = 1.5$ leads to an increase of about 35%. Unlike R_g, polydispersity has a small influence on the value of D_0. We should note that Eqs. (1)–(5) do not consider out-of-plane bending deformations.

3.   Experimental Techniques

3.1.   *Dynamic light scattering (DLS)*

The DLS experiment records the autocorrelation function $G(q,t) = \langle I(q,t)I(q,0)\rangle/\langle|I(q,0)|^2\rangle$ of the light scattering intensity $I(q,t)$ at a scattering wave vector $q = [(4\pi n_s/\lambda_0)sin(\theta/2)]$ with λ_0, θ and n_s being, respectively, the laser wavelength in vacuum, the scattering angle (between incident and scattered light) and the refractive index of the solvent (for dilute solutions). The desired electric field correlation function is $C(q,t) = a(q)g(q,t)$, where the normalized $g(q,t) = [G(t)-1]^{1/2}/a(q)$ and $a(q)$ is the amplitude of $G(q,t)$ at the shortest time ($\sim 0.1\ \mu s$) of the ALV-5000/E correlator. The decay rate Γ ($= 1/\tau$, the relaxation time) and the amplitude $a(q)$ of the relaxation process are obtained[1] from the inverse-Laplace transformation (ILT) of $C(q,t)$. If one relaxation process contributes to $C(q,t)$, the full scattered intensity by the solution, $I(q)$, at a given q can be used to compute the excess Rayleigh ratio $R(q) = [(I(q) - I_s)/I_T]R_T(n_s/n_T)^2$, where T stands for the standard toluene and I_s is the (q-independent) light scattering intensity of the solvent. For more than one relaxation process in $C(q,t)$, the time average total light scattering intensity, $\langle I(q)\rangle$, measured by static light scattering (SLS), includes more than one contributions and hence cannot be used to compute $R(q)$. Instead, $(q) = a(q)\langle I(q)\rangle$, with $\langle I(q)\rangle$, where $a(q)$ accounts for the desired amplitude accessible only by DLS. $R(q)$, which is proportional to the form factor, is related to the weight-average molar mass (M_w), and the characteristic size of the scattered moiety. In the case of Gaussian coils with gyration radius

(R_g) and in the low qR_g ($\ll 1$) range,

$$\frac{K \cdot c}{R_q} = \frac{1}{M_w}(1 + 2M_w A_2 c)\left(1 + \frac{q^2 R_g^2}{3}\right) \tag{6}$$

where c is the solute concentration of, A_2 the second virial coefficient and $K = [2\pi n_S(dn/dc)]^2/(N_A \lambda_0^4)$ is the optical contrast with N_A being the Avogadro's number and dn/dc the refractive index increment in the solvent.

3.2. *Fluorescence correlation spectroscopy (FCS)*

The method of FCS utilizes fluorescent intensity fluctuations $\delta I_F(t)$ through the focal spot, an ellipsoidally shaped diffraction-limited spot of the Gaussian illumination beam profile. Such fluctuations, which are recorded when a molecule crosses the focal spot with axial size $2z_0$ and lateral size $2w_0 \sim$ 300 nm ($s = z_0/w_0$ being the aspect ratio), stem from spontaneous spatiotemporal fluctuations in the number density of fluorescently labeled molecules simultaneously present in the spot at a given instant. These fluorescence intensity fluctuations, $\delta I_F(t)$, with respect to the average fluorescent intensity, $I_F(t)$, define the experimentally recorded fluorescent intensity autocorrelation function, $G(t) = \langle \delta I_F(t) \cdot \delta I_F(t+\tau) \rangle / \langle I_F(t) \rangle^2$. The temporal resolution of the FCS method spans between 100 ns–5 s. A monochromatic laser light source ($\lambda = 633$ nm) is used to excite fluorescently labeled polymers, while the emitted fluorescent signal is collected by the very same optical setup (inverted Carl Zeiss microscope with Confocor 2 Module) which was used for excitation. The experimental $G(t)$ for single translational diffusion is represented[18] by,

$$G(t) = 1 + N^{-1}\{[1 + T/(1-T)e^{-t/\tau_T}][(1 + t/\tau_D)^{-1}] \\ \cdot [1 + t/(s^2 \tau_D)^{-1}]\}^{-1/2} \tag{7}$$

where T and τ_T are the fraction and decay time of the triplet state, N represents the average the number of diffusing fluorescent species through the focal spot and τ is the corresponding the lateral diffusion time with diffusion coefficient $D = w_0^2/4\tau_D$.

4. Polymers and Supramolecular Assemblies

This overview of rigid polymers is not exhaustive, it is rather intended to present some examples of persistent polymers, in order to stress the correlation between structure and persistence length. We confine the presentation to cases where reliable data on persistence length have been reported.

4.1. *Persistent covalent polymers*

Rigid-polymer backbone and bulky monomer structure

The polymer backbone can greatly influence the chain rigidity (Fig. 1). Polymers with highly conjugated or sterically hindered backbones, such as PPV (1),[5] have longer persistence lengths in solution (6–40 nm) than those with backbones of aliphatic chains, such as polystyrene (3) (1 nm) (Table 1).[19] The persistence length of PPV was indirectly inferred from photoluminescence spectra. Also polymers with a PPP (2) backbone are considered as typical rigid-rod polymers due to the linear chain of stiff phenylene units (Table 1).

Flexible side chains attached to the backbone are necessary to improve solubility and avoid aggregation, as flexible side chains act like a polymer bound solvent shell, disguise the backbone elements and help to achieve processability.[3] However, such chemical functionalization leads to an increased torsion of the polymer backbone that decreases the conjugation length. The synthesis of PPPs with a large variety of side groups has been pioneered by G. Wegner and coworkers.[4] Well-defined single-stranded PPPs with good solubility and strong shape anisotropy have a large persistence length. For example, "hairy rod" PPPs, PPPs with aromatic sulfonate side groups) (4) have a persistence length of 25 nm.[20]

Various methods have been developed to overcome the diminishing solubility of unfunctionalized PPPs with increasing number of repeating units. For instance alkyl chains can be introduced and increase the solubility.[21] Another method is to prepare soluble precursors which can be deposited and converted to PPP on the surface. One example for this method is the 3,6-connected cyclohexadiene oligomers (5) (inset to Fig. 2). The rigid linear structure was anticipated due to the trans-configuration of the monomeric units.[22] As R(q)/c (Eq. (6)) is q-independent for these molecules, information on the size and shape can only be deduced from the translational diffusion D

Table 1: Molecular structure of PPV (1), PPP (2) and polystyrene (PS) (3) along the reported persistence lengths.[5,19,20]

PPV (1)	PPP (2)	PS (3)
6 – 40 nm	6 – 25 nm	1 nm

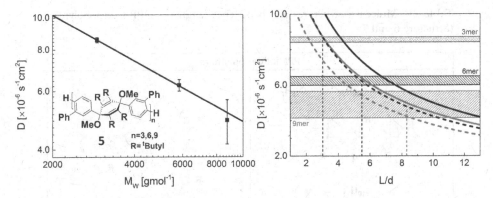

Figure 2: Left panel: Molecular weight dependence of the translational diffusion D for three 3,6-connected cyclohexadiene oligomers (**5**) (inset). The solid line represents a scaling relation $D \approx M_w^{-0.5}$. Right panel: Computed translational diffusion D versus aspect ratio L/d (Eq. (8)) for cylinder or ellipsoid with length, L (long axis) and diameter d (short axis). The lines are theoretical predictions for $d = 1.1$ nm (solid lines) and $d = 1.3$ nm (dashed lines) for rod-like shape (red lines) and for prolate ellipsoid-like shape (black lines). The horizontally shaded regions denote the range of experimental D for the three indicated oligomer solutions and the three dashed vertical lines indicate the proposed aspect ratios.

in dilute diethyl ether solutions measured by DLS. As shown in Fig. 2, the dependence of D on molar mass M_w indicates a scaling, $D \approx M_w^{-0.5}$ which is reminiscent of Gaussian coils. This is, however, a fortuitous observation and is due to the different aspect ratio L/d, since D is well represented by either a rod-like or prolate ellipsoid shape,

$$D = [kT/(3\pi\eta_s L)] \times T(L/d) \tag{8}$$

with length L (long axis), diameter, d (short axis) that defines the function $T(L/d)$.[22] Assuming $d = 1.2 \pm 0.1$ nm, the L values (3.5 ± 01 nm, 6.6 ± 0.5 nm and 9.7 ± 2.5 nm) conform to the trimer, hexamer and nonamer ratio 1:2:3. Thus the persistence length should be larger than 9.7 nm, the nonamer length, and its determination requires the synthesis of longer 3,6-connected cyclohexadiene chains.

To get a deeper understanding of persistent polymers, synthesis of polymers with different monomer architecture would help to define general trends. Therefore, PPP ladder polymers with different side chains were synthesized (Table 2, structure **6** and **7**).[23] Despite the double-stranded molecular structure in which rotation cannot occur, this new polymer **6** is characterized by a rather short persistence length (7 nm) as estimated by PCS and PFG-NMR using $d = 1.8$ nm.[24] In comparison to PPP ladder polymers, hairy rod PPPs have a persistence length of 25 nm. However, PPP ladder polymers with bulky side

Table 2:   Molecular structures and persistence lengths of PPP ladder polymers (Structure 6 and 7).

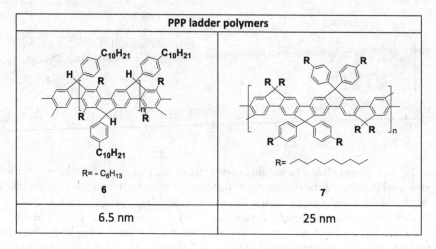

PPP ladder polymers	
6	7
6.5 nm	25 nm

groups (**7**) have also a persistence length of 25 nm as measured by DLS (Table 2, Structure 2).[25] This means that the introduction of bulky side groups in PPP ladder polymers increases the persistence length.

To sted more light on the relationship of polymer chain architecture and persistence length of rigid polymers, [poly(9,9-bis(2-ethylhexyl)fluorene] (**8**), another class of conjugated polymers, was investigated. In contrast to PPP ladder polymers (structure **6** and **7**), in polyfluorene rotation of the repeating units about interring bonds affects the overall shape of the molecule. Polyfluorenes are important materials since they find applications in polymer-based light-emitting diodes. SANS and translational diffusion measurements in dilute solutions revealed single polymers with a persistence length of 7.0 ± 0.5 nm and a cross-section diameter of 1.8 ± 0.5 nm.[26] For a dioctyl-substituted poly(fluorene) (**9**), a persistence length of 8.0 ± 1.0 nm was reported based only on R_g data.[27] The difference in the persistence length between PPPs and polyfluorenes could be explained with the different bond angles of the adjacent monomer units (0° and 21° for PPPs and PF, respectively).[26]

Combination of rigid polymer backbones and supramolecular organization

Besides the introduction of hydrophobic chains in PPP, PPP can be functionalized with hydrophilic side chains such as poly(ethylene glycol) leading to amphiphilic molecules (**10**). Fütterer *et al.* introduced polymers functionalized with triethylene glycol chains and dodecyl chains in each repeating unit.

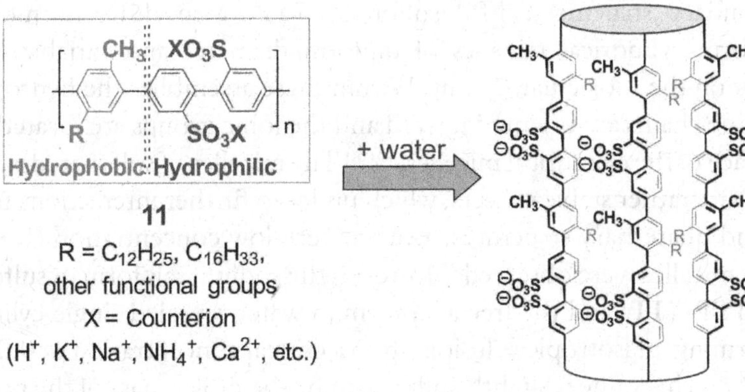

Figure 3: Molecular structure of sulfonated rigid rod PPPs (**11**) and schematic representation of the hierarchical structure formation of sulfonated rigid-rod PPPs (**11**) in aqueous solution.

In contrast to linear, amphiphilic block copolymers, this substitution scheme leads to amphiphilic polymers in which the boundary between the hydrophobic and hydrophilic segments is parallel to the polymer backbone. The polymer can be dissolved in water by comicellization with low-molecular weight nonionic surfactants such as tetraethylene glycol monooctyl ether. Cryo-TEM reveals that the amphiphilic PPP in the micellar solution forms fiber-like assemblies with a contour length greater than 200 nm and a diameter of 5.9 nm in the presence of an excess of surfactant micelles, with a persistence length nearly equal to the contour length. The hydrophilic segments point to the solvent and the hydrophobic part points inwards leading to cylindrical micelles.[28]

Amphiphilic rigid polymers can not only be synthesized with neutral, polar side groups, but also by the introduction of electrolyte functions. Charged side chains as it is the case in polyelectrolytes can lead to chain stiffening due to ionic interactions.[29] The persistence length of sulfonated rigid rod PPPs (**11**) was measured by DLS (Fig. 3). For experiments on individual polyelectrolyte chains, it is important that characterizations of polyelectrolytes are not straightforward due to strong tendency for aggregations even at very low concentrations. Rigid polyelectrolytes exhibit limited solubility due to aggregation effects, despite the presence of side chains.[30–32]

The dependence of the intrinsic viscosity on the molar mass was determined from both the Mark–Houwink–Sakurada equation and the wormlike chain model. The low value of the persistence length (13 nm) indicates that these polymers are relatively flexible.[33,34] The macromolecule consists of a hydrophilic and a hydrophobic part which self-assembles to well-defined cylindrical micelles of a number of PPP chains oriented parallel to the cylinder axis (secondary structure). PPP sulfonates (**11**) with different molecular weights form cylindrical micelles of uniform diameter and variable length, depending on the molecular weight. Within these assemblies the hydrophobic aliphatic side chains are oriented inward and the ionic groups are located at the outer surface of the cylindrical micelles.[35,36] The micelles of sulfonated rigid rod PPPs (**11**) are rather stable objects, which undergo further interactions to form tertiary and quaternary structures. Even at very low concentration (8×10^{-4} g/L) only micelles were observed. However, the sodium salt form of sulfonated rigid rod PPPs (**11**) and the free acid form in water revealed single cylindrical objects forming an isotropic solution up to a critical concentration of 0.02 g/L at 20°C.[37,38] This value is slightly influenced by the molar mass of the polymer, the counterion species and the temperature. Also positively charged conjugated polymers, e.g. a trimethylammonium-hexyl functionalized polyfluorene (**12**), self-assemble to cylindrical assemblies in strong polar solvents.[39]

Lateral expansion of the rigid polymer backbone

The polyphenylene backbone can also be expanded into a second dimension. Here we present a polyphenylene backbone grafted with oligophenyl groups (structure **13** and **14**, inset to Fig. 4). In contrast to polyphenylene with no or small linear side chains, the rotation of the benzene rings in the backbone is blocked due to the twisted rings in the side groups. Polyphenylene (**13** and **14**) were synthesized by way of repetitive Diels-Alder reactions and the persistence length was measured by PCS. For dilute polymer solutions, $C(q, t)$ arises from thermal concentration fluctuations with correlation length $\sim 2\pi/q$ and a single diffusive (q^2-dependent) relaxation rate $\Gamma(q)$ as depicted in Fig. 4 for the case of a polyphenylene solution (1.5 g/L) in THF. $C(q, t)$ slows down with decreasing q, the q-dependent R_q in the plot of the upper inset to Fig. 4 yields $M_w = 4.7 \times 10^5$ gmol^{-1} and $R_g = 37$ nm (Eq. (3)). The relaxation rate $\Gamma(q)$ obtained from the virtually exponential shape of $C(q, t)$ leads to the translational diffusion of the polymer chain, $D(q) = \Gamma(q)/q^2$ (lower inset to Fig. 4),

$$D(q) = D_0(1 + Cq^2 R_g^2), \quad D_0 = kT/(6\pi\eta_s R_h) \qquad (9)$$

where k, T and η_s are the Boltzmann constant, the absolute temperature and the solvent viscosity, and D_0 is the translation diffusion coefficient with hydrodynamic radius R_h ($= 29$ nm) and the slope C (~ 0.17) depends on the polymer shape; $R_g/R_h \sim 1.3$, typical value for coils with $C \sim 0.18$.[40]

For two well-characterized samples with molecular weight 108 kDa and 470 kDa (Fig. 4) corresponding to contour length $L = 110$ nm and 480 nm,

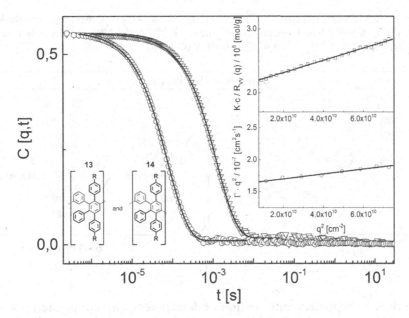

Figure 4: Relaxation functions $C(q, t)$ for a dilute solution of polyphenylene **13** and **14** (lower inset) in THF at two scattering wave vectors ($q = 0.0072$ nm^{-1}, $q = 0.027$ nm^{-1}) at 20°C. Insets: The normalized light scattering intensity (Eq. (5)) (upper) and the translational relaxation rate (lower) obtained from the ILT of $C(q, t)$.[41]

respectively, the diffusion D_0 (Eq. (2)) and R_g (Eq. (4)) can be well represented by persistence length, $l_p \sim 9$ nm. Dependent on the molecular weight, these polyphenylene chains behave either as semi-rigid rod ($L/l_p \sim 12$ precursor, non-planarized) or as semi-flexible ($L/l_p \sim 52$) polymers.[41] Note that a unique characterization (absolute molecular weight, R_g and D_0) of at least two different polymer fractions is necessary to correctly assign the chain conformation.

Dendronized polymers

Side chains can have a significant influence on the rigidity of a polymer chain. For example, Frederickson predicted a stiffening of the main chain for bottlebrush polymers as a consequence of the presence of a dense corona of side chains.[42] Here the question arises: Does the introduction of bulkier groups make the polymer even stiffer than with linear side chains? Progress in synthetic polymer chemistry allows for the introduction of not only linear side chains but also bulky side groups. Schlüter *et al.* introduced dendronized polymers (denpols). These are polymers with multiple dendritically branched lateral chains (dendrons) which are covalently linked at regular intervals.[43]

Table 3: Molecular structure of polyfluorene with bulky polyphenyene dendrimer groups (**16**), denpols based on poly(norbornene) (PNB-G_3) (**17**) and denpols based on poly(endo-tricycle[4.2.2.0]deca-3,9-diene) (PTD-G_3) (**18**).

Dendronized polymers		
16	PNB-G_3 **17**	PTD-G_3 **18**
-		6 - 8 nm

In the first reported combination of dendrimers with conjugated polymers in which Fréchet dendrons were grafted on PPP (**15**) the sterically demanding dendritic side chains led to strong torsion about the phenylene-phenylene bonds.[44] In contrast to that, a polyfluorene with bulky polyphenyene dendrimer groups has been prepared (Table 3, structure **16**). The absorption and emission spectra of these molecules indicate that the bulky dendrimer side chains do not cause extra torsion between the fluorene units.[45]

Other examples are denpols based on poly(norbornene) (PNB) (**17**) and poly(endo-tricycle[4.2.2.0]deca-3,9-diene) (PTD) (**18**) backbones (Table 3). Their synthesis by ring-opening metathesis polymerization (ROMP) led to fully grafted and high molecular weight denpols with narrow polydispersity.[34] Poly(norbornene) containing a small alkyl chain as a solubilizing group (called "precursor") was prepared as a reference material for comparison with the denpols.

Typical DLS results of the "precursor" of the dendonized PNB-G3 PNB in dilute solution in chloroform are shown in Fig. 5. The relaxation function $C(q,t)$ for the "precursor" is represented by a single relaxation process with a narrow distribution of relaxation time as seen in Fig. 5 at all examined q's in the dilute regime. The diffusion coefficient D of the chains with $qR_g < 1$ is expectedly (Eq. (9)) independent of the wave vector q (no contribution of internal chain dynamics in Eq. (9)) as shown in the upper inset to Fig. 5. Hence, $D = D_0$ is attributed to the center-of-mass motion leading to the hydrodynamic radius R_h (Eq. (9)). The average scattered intensity expressed as the polarized light scattering ratio Kc/R_q increases linearly with q^2. According to Eq. (5) the

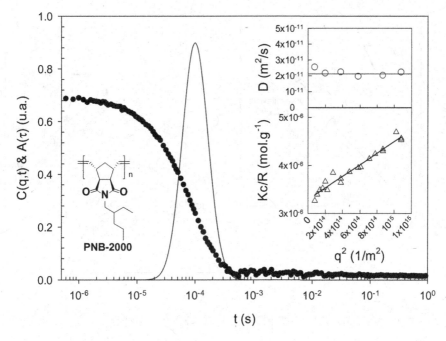

Figure 5: Relaxation function for the concentration fluctuation in PNB-2000/chloroform dilute solution ($c = 1.22$ g/L) at $q = 2.41 \cdot 10^{-2}$ nm^{-1} along with the relaxation time distribution function $A(\tau)$ obtained from the ILT of $C(q, t)$ (solid line). Inset: Translational diffusion coefficient (upper) and the reciprocal reduced intensity, Kc/R (lower panel) represented by Eq. (6).[46]

apparent molar mass and radius of gyration are computed from the intercept and the slope of the solid line (lower inset to Fig. 5) respectively. For the denpol based on poly(norbornene), the concentration of Kc/R_q yields the molar mass M_w ($= 606$ kDa), R_g (46 nm) and R_h ($= 23$ nm) at $c = 0$, and the second virial coefficient, A_2 ($= 7.7 \times 10^{-4}$ mol·cm^3/g^2). Again, the full molecular characterization of both precursor and dendronized polymers **17** and **18** is prerequisite of a unique estimation of the chain persistency.

As the probed q-range (lower inset to Fig. 5) of the precursor PNB falls in the low-q limit ($q \cdot l_p \ll 1$), a direct access of the persistence length l_p from the scattering pattern $R(q)$ is ambiguous. The persistence length can nonetheless be estimated independently from the radius of gyration R_g wherever available (Fig. 6(a)) and also from the diffusion coefficient D_0 (Fig. 6(b)). The mean-square-radius of gyration of monodisperse semi-flexible polymers is a function of the persistence length l_p and the contour length L, as given by Eq. (1) and Eq. (4) for monodisperse and polydisperse semi-flexible polymers, respectively. If L and R_g are known, Eq. (4) allows the measurement of l_p.

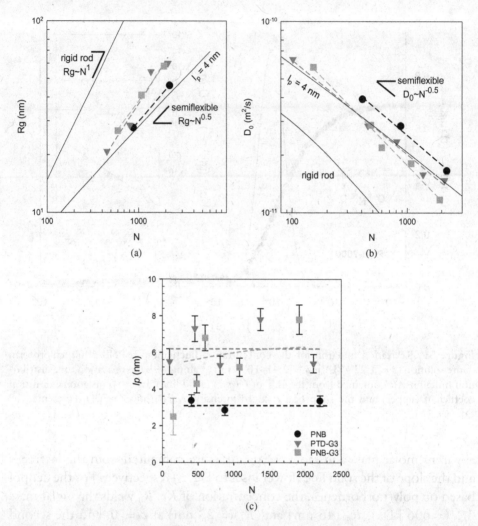

Figure 6: Radius of gyration (a) and diffusion coefficient (b) as function of N (number of monomer per chain) for PNB (filled circle), PNB-G3 (17) (filled square), and PTD-G3 (18) (reverse triangle) in log/log plot (dash line is a guide for the eyes). Solid lines represent $R_g(N)$ from the rigid rod ($R_g^2 = L^2/12 + R_c^2/2$ with cylinder radius, $R_c = 1$ nm), and semi-flexible model (Eqs. (1) and (4)) in (a) and $D_0(N)$ from rigid rod and semi-flexible model (Eq. (2)) in (b). Persistence length estimated from Eq. (2) as a function of polymer size for PNB ($\bullet$), PNB-G3 (17) (green square), and PTD-G3 (18) (red reverse triangle) (dash line is a guide for eyes).[46]

Figure 6(a) shows the variation of the experimental R_g with the degree of polymerization N along with the theoretical prediction for a rigid rod ($l_p > L$) shape. The conformation of these denpols **17** and **18** containing third generation ester dendron is far from a rigid rod behavior, assuming much smaller R_g than expected for a rod-like structure with the same N. Instead, these denpols resemble semi-flexible chain behavior with R_g obtained from Eq. (4). The second solid line in Fig. 6(a) represents $R_g(N)$ of monodisperse semi-flexible chains with $l_p = 4$ nm conforming to $R_g \sim N^{0.5}$ for $L \gg l_p$. This picture is also confirmed by the experimental diffusion coefficients shown in Figure 6(b) for all examined polymers as a function of N. Since Yamakawa-Fujii (YF) theory of wormlike chain[15,16] YF is valid for $L/d > 50$, we employed Eq. (2) which is a more general expression of D_0 for semi-flexible chains[17] covering larger L/l_p range. The hydrodynamic d was fixed to 2 nm in order to consistently compute l_p for all dendronized polymers (Fig. 6(b)). The solid line in Fig. 6(b) represents the prediction of Eq. (2) for the indicated l_p (= 4 nm) value. The persistence length obtained from R_g is systematically larger than the one obtained from D_0.

All the measured polymers (including the precursor) can be considered as semi-flexible, since their l_p falls between that of flexible polymers like polystyrene in theta solvent ($l_p \approx 1$ nm) and for semi-rigid polymer like polyfluorene in toluene ($l_p \approx 7$ nm).[26] For the dendronized polymers, l_p exceeds the value for the "precursor" chain as can be seen in Fig. 6. This is the expected effect of the steric hindrance induced by the lateral dendrimers along the polymeric backbone. The conformation of denpols with a third generation side dendron conforms to a semi-flexible chain with a persistence length of about 6–8 nm as characterized by static and dynamic light scattering. The assumption of extremely high chain rigidity for this polymer is not supported.[46] However, we should mention that not only does the steric hindrance of the dendron has an influence on the persistence length of the dendronized polymer, but also the solvent quality and the diameter of the molecule can matter.[47] In contrast to dendronized polymers, a combination of molecular dynamics simulations and analytical calculations for bottle brush polymers, showed, that the Kuhn length l_k increases with increasing the side-chain degree of polymerization n_{SC} as $l_k \sim n_{SC}^{0.46}$.[48] Theoretical studies on the chain conformation of dendronized polymers also lead to much debate about the relationship between structure and the persistence length.[46,49]

Helical chains

There are a lot of known helical polymers. For example, polyisocyanates (**19**)[50,51] (Table 4) have a persistence length of 37 nm to 50 nm and helical poly(γ-benzyl-L-glutamate) (**20**) have a persistence length up to approximately 200 nm at high molecular weights.[29]

An interesting consequence of persistence is the translation-dynamic coupling both which appears in helical polymers (Table 4) and also in

Table 4: Molecular structure and persistence length of polyisocyanate (**19**) and poly(γ-benzyl-L-glutamate) (**20**).

Polyisocyanate	Poly(γ-benzyl-L-glutamate)
19	**20**
37 – 50 nm	200 nm

poly(p-phenylene)s (**2**) (Table 1) and PPP ladder polymers (structure **7** and **8**) (Table 2).[25,50] The collective orientation dynamics were found to become heterogeneous with increasing concentration in the semi-dilute regime (above the overlap concentration). The relaxation function of the orientational order parameter fluctuations recorded by depolarized DLS displays a bimodal shape with a peculiar q-dependence as shown in Fig. 7(a) for a semi-dilute solution of polyfluorene in toluene at different scattering angles. The dynamics counterintuitively slow down with increasing scattering angle, i.e., q, being much faster in the forward ($\theta = 0°$) than in the backward ($\theta = 180°$) scattering (Fig. 7(b)). This apparently intriguing evolution can be understood by employing,

$$C(q,t) = f_f(t)cos^2(\theta/2) + f_b(t)sin^2(\theta/2) \tag{10}$$

where $f_f(t), f_b(t)$ are the forward and backward components, respectively, and for isotropic solutions with random orientation, $f_f(t) = f_b(t)$. The observed differentiation can be attributed to non-random spatial orientation correlations due to rotation-translation coupling proposed for small molecules.[52–56] For semi-flexible polymers, however, the effect is strong and the dynamics are slow that renders the finding unique. A further advantage is the observation of the effect at equilibrium without macroscopic flow. The coupling between shear modes and orientation, $R = 1 - \tau_f/\tau_b$ with τ_f, τ_b presenting the fast (in forward) and slow (in backward scattering) relaxation times, is system dependent. The pure orientation dynamics (slow component in Eq. (7)) has a non-exponential shape due to the various rotational times and length polydispersity. Studies of the orientation dynamics of semi-flexible polymers are very few so far but promising as they open a new route to elucidate the role of the inherent persistence length and the anisotropic interactions

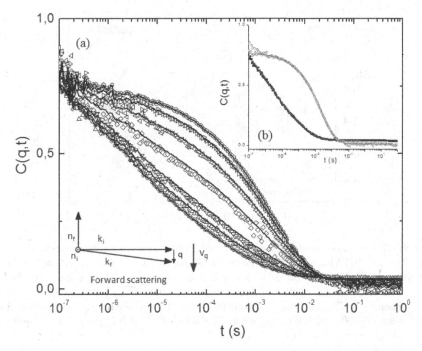

Figure 7: Experimental orientation relaxation functions $C(q, t)$ at different scattering angles (θ) from $15°$ to $150°$ (from *left* to *right*) for a 33.34 wt% [poly(9,9-bis(2-ethylhexyl)fluorene] (7) (89 kDa) solution in toluene at $20°$C. The *solid lines* denote the representation of the experimental functions by Eq. (10) without adjustable parameters using $C(q, t)$ at the extreme angles (top inset: solid and open symbols for forward and backward scattering) as reference. Bottom inset: The shear flow velocity V_q effectively couples to translation in the forward scattering.

emerging with increasing concentration to the flow properties. Further new experiments on concentrated solutions of polymers with large persistence length e.g. substituted phthalocyaninato-polysiloxane[57] will complete the necessary phenomenology necessary to develop a molecular theory.

By changing the pH or the temperature, polypeptides can undergo a helix-coil transition.[58] It still remains a major challenge to macromolecular chemistry to design polymers which would undergo such transitions and would yield identical fold structures every time the transition is made.

4.2. *Supramolecular assemblies*

We compare the persistence length of different amphiphiles consisting of a hydrophobic part (either alkenyl or benzene) and a hydrophilic part (amide bonds and/or poly(ethylene glycol)). Therefore we chose a low molecular weight benzene derivative functionalized with amides and triethylene glycol.

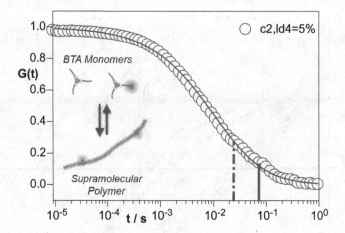

Figure 8: Fluorescence intensity correlation function $G(t)$ for 1,3,5-benzenetricarboxamide (BTA) (**21**) with [BTA] = 5 μM label fraction = 5% (open blue circles) represented by Eq. (7) (main manuscript). The vertical dashed-dotted line denotes the effective internal relaxation time τ_c, whereas the solid vertical line indicates the computed τ_d based solely on center of mass diffusion ($D = 7 \times 10^{-9}$ cm^2/s from DLS) scaled down for translational diffusion over the FCS waist ($\sim$300nm). Inset: Supramolecular BTA fibers with large contour length ($L \sim$ 5 μm) undergo continuous dynamic exchange with their — unlabeled and labeled — repeat units/monomers.[59,60]

Meijer *et al.* studied 1,3,5-benzenetricarboxamides (BTAs) (**21**) (Fig. 8) as building block of one-dimensional fibers. The monomers are well-known to self-assemble into fibers via threefold hydrogen bonding, due to the amide and stacking of the BTA cores in a columnar helical fashion. In this case one molecule is in one layer. The persistence length of such fibers is reported by small X-ray scattering to be of several hundreds of nanometers.[6] In contrast to the covalent polymers, the ability of many directional interactions affect the conformation energy landscape in Fig. 1 and thereby bias trans conformers. In addition, out-of-plane bending deformation is a conceivable contribution to the persistence length.

However, both an extension to lower q's and access to D_0 are necessary to support long l_p values. DLS experiments[60] indicate high degree of association of the long-lived supramolecular BTA chains and long contour lengths ($\sim$few μm) in agreement with fluorescence confocal imaging.[59] Based on the low D_0 values obtained from both DLS and FCS (Fig. 3), l_p is indeed large (of the order of 150 nm). According to Fig. 8, the fluorescence intensity correlation function $G(t)$ reveals faster dynamics (dashed line) than DLS (vertical solid line) even for heavily labelled (5%) BTA fibers. This discrepancy can be rationalized by the additional contribution of fast internal dynamics to $G(t)$ as the longer fibers exceed the FCS observation spot in agreement with the semirigid shape of the BTA fibers.[60]

Prominent examples in nature for rod-like self-assembled structures are the Tobacco Mosaic Virus and the fd viruses. The Tobacco Mosaic Virus and the fd viruses were the first model systems for rod-like structures.[61,62] Percec *et al.* used this virus as model system and synthesized molecules with crown ethers, and hydrogen bonding parts. These molecules self-assembled to cylindrical self-assembled structures which are similar to the self-assembly of the proteins in the tobacco mosaic virus.[63,64] One of the most ubiquitous self-assembly processes in nature with high persistence lengths is the hierarchical organization of peptides into long filaments, bundles and networks. Synthetic oligomeric peptides can be designed which undergo one-dimensional self-assembly in solution to form β sheet tapes a single molecule in thickness and micrometers in length. The tapes behave like semi-flexible chains with persistence length of several hundred nanometers and much longer contour lengths even at very low concentrations ($c = 0.1$ nM).[65] They have an anisotropic rigidity like the ladder polymers with mirror symmetry.[10–12]

Using different block copolymers, Eisenberg *et al.* obtained spherical and cylindrical micelles or vesicles in solution.[66] According to the model of Israelachvili,[67] which is strictly speaking only valid for low molecular surfactants, with geometric considerations the self-assembly can be determined dependent on the shape of the molecule. Eisenberg and Discher extended the concept of Israelachvili to polymers.[68] In this concept the packing parameter p is defined with the following equation: $p = v/(la)$. a is the surface of the polar head group, v is the volume of the alkyl chains and l is the length of the alkyl chains. For $0 < p < 1/3$, spherical micelles can be predicted and for $1/3 < p < 1/2$ worm-like superstructures can be formed. A high molecular weight poly(ethylene oxide-*b*-butadiene) (**22**) self-assembles in water and forms cylindrical worm-like micelles. In contrast to the self-assembly of the BTA (**21**), poly(ethylene oxide-*b*-butadiene) (**22**) places several polymers in one layer. The micelles, after crosslinking of the butadiene core, exhibit unusual linear and nonlinear flow properties that presumably reflect the micelle stiffening upon cross-linking.[7]

To compare the low-molecular weight amphiphiles (e.g. **21**) and high-molecular weight amphiphiles (e.g. **22**), we have synthesized and studied amphiphilic compounds based on hexaphenylbenzene (HPB) (**23** and **24**), with two different PEG substitution patterns and investigated their self-assembly behavior in water by cryo-TEM and DLS (Fig. 9). These molecules were studied because of the pronounced π-π interactions of

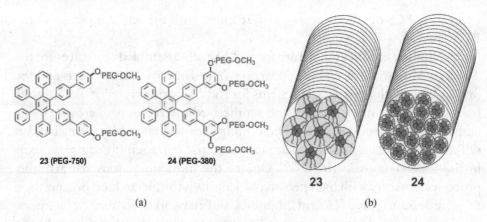

(a)   (b)

Figure 9:   (a) Molecular structure of HPB-PEG derivatives. (b) Schematic representation of the proposed self-assembled equilibrium structures of **23** and **24**.

hexaphenylbenzene. HPB-PEG derivatives form bundles of hydrogel fibers in very dilute aqueous solution. These self-assembled structures can be represented by a worm-like model with three adjustable parameters, L_w, Kuhn segment, l_k ($l_k = 2l_p$, the persistence length), and thickness, d assuming sticky hydrodynamic boundary conditions. The overall diameter of the bundles of **23** and **24** is in both cases 30 nm. The persistence length of **23** and **24** was 30 nm and 40 nm, respectively.[8]

Surprisingly, the persistence length of the assemblies appears to be almost unaffected by the number of molecules per layer which is much higher for **24**. It was concluded that the persistence length is controlled less by the PEG chains than by the columnar packing.

5.   Concluding Remarks — Perspectives

This chapter has documented some examples of the recent literature on the persistence of polymers and supramolecular assemblies obtained mainly by scattering techniques. The chain rigidity originates either from a rigid aromatic backbone, winding of chains into helices or by grafting of side chains of various architectures (linear, branched) onto the main chain. We can classify different types of polymers and supramolecular assemblies for which the persistence length is reliably measured (Table 5). There is, however, no clear structure-rigidity relationship such as to enable a reliable prediction of the persistence length from the monomer structure as observed in the case of the dendronized polymers. Nevertheless, supramolecular assemblies assume distinctly larger

Table 5: Overview of persistent polymers and supramolecular assemblies with conformation and persistence lengths.

Polymer	Conformation	Persistence length	References
PPV (**1**)	Semi-flexible-Rod	6–40 nm	5
Polystyrene (**3**)	Coil	1 nm	19
Hairy rod PPPs (**4**)	Rod	25 nm	20
PPP ladder polymers (**6,7**)	Semi-flexible-Rod	6.5–25 nm	24,25
Polyfluorene (**8,9**)	Semi-flexible	7–8 nm	26,27
Sulfonated PPPs (**11**)	Semi-flexible	13 nm	33,34
Polyphenylene backbone grafted with oligophenyl groups (**13,14**)	Semi-flexible	9 nm	41
PNB-G3 (**17**), PTD-G3 (**18**)	Semi-flexible	6–8 nm	46
Polyisocyanate (**19**)	Helical	37–50 nm	50,51
Poly(γ-benzyl-L-glutamate) (**20**)	Helical	200 nm	29
Supramolecular polymer consisting of BTA (**21**)	Rod	Several hundreds of nanometers	6,59,60
Oligomeric peptides	β sheet tapes	Several 100 nm	65
HPB-PEG derivatives (**23,24**)	Bundles of fibers	30–40 nm	8

rigidity than covalently bonded polymers while chain rigidity affects the flow behavior and dynamics in the non-dilute concentration regime.

Depending on the intermolecular interactions, single polymers can also form supramolecular assemblies with cylindrical symmetry rationalized by the packing parameter. However, there are many examples for which the persistence length of the formed supramolecular assemblies cannot be reliably predicted and the design principles remain empirical.

Simulations[69] of the chain persistence depending on the conformational potential (Fig. 1) in theta solvents and the configuration of the neighboring repeating units (for example inset to Fig. 2) followed by comparison with the experiment are vital to unravel the particular monomer structure- conformation relationship. To gain a better understanding of the correlation between expanding the conjugation in the third dimension and the resulting persistence length, for example, the synthesis of oligomers (**26,27**) and polymers (**28**) of PPPs from sterically π-congested 2,2′,6,6′-tetraphenyl-1,1′-biphenyl (**25**) units was performed (Fig. 10).

Investigation of the self-assembly revealed significant π-orbital interactions along the para-phenylene backbone of the π-congested systems together with through-space conjugation among the π-orbitals of the stacked peripheral phenyl rings led to an increased effective conjugation length.[70]

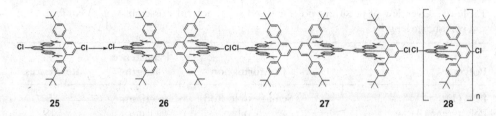

Figure 10: Synthesis of oligomers **26,27** and polymer **28**.[70]

In the case of supramolecular assemblies forming either from small molecules (Figs. 8, 9) or macromolecules (Fig. 3), the situation is more complex, as monomer structure, antagonistic (amphiphilic, electrostatic) and solvent mediated interactions, and dynamics can bias the self-assembly involving metastability in the structure formation. Whether and how this complexity would be manifested in the structure (static) persistence is an open big question given the importance to biological systems. Finally, it is important to consider the transition from the solution to the solid state and to compare covalent polymers with supramolecular assemblies. Based on the rich rheological behavior of supramolecular polymers[71] and supramolecular assemblies[72] in the non-dilute regime missing studies over wide concentration range are considered necessary.

Acknowledgments

G.F. thanks Prof. G. Wegner for fruitful discussions.

References

1. H. Frauenrath, Dendronized polymers-building a new bridge from molecules to nanoscopic objects. *Prog. Polym. Sci.*, **30**(3–4), 325–384 (2005).
2. A. Aggeli, M. Bell, N. Boden, J. N. Keen, P. F. Knowles, T. C. B. McLeish, M. Pitkeathly, S. E. Radford, Responsive gels formed by the spontaneous self-assembly of peptides into polymer β-sheet tapes. *Nature*, **386**, 259–262 (1997).
3. G. Wegner, Shape persistence as a concept in the design of macromolecular architectures. *Macromol. Symp.*, **201**, 1–9 (2003).
4. M. Rehahn, A.-D. Schlüter, G. Wegner, Soluble poly(para-phenylenes) 2. improved synthesis of poly(para-2,5-di-n-hexylphenylene) via Pd-catalysed coupling of 4-bromo-2,5-di-n-hexalbenzeneboronic acid. *Polymer*, **30**(6), 1060–1062 (1989).
5. C. L. Gettinger, A. J. Heeger, J. M. Drake, D. J. Pine, A photoluminescence study of poly(phenylene vinylene) derivatives — the effect of intrinsic persistence length. *J. Chem. Phys.*, **101**(2), 1673–1678 (1994).

6. M. W. Baker, L. Albertazzi, I. K. Voets, C. M. A. Leenders, A. R. A. Palmans, G. M. Pavan, E. W. Meijer, Consequences of chirality on the dynamics of a water-soluble supramolecular polymer. *Nature Commun.*, **6**, 1–12 (2015).

7. Y.-Y. Won, K. Paso, H. T. Davis, F. S. Bates, Comparison of original and cross-linked wormlike micelles of poly(ethylene oxide-*b*-butadiene) in water: Rheological properties and effects of poly(ethylene oxide) addition. *J. Phys. Chem. B*, **105**(35), 8302–8311 (2001).

8. K. Wunderlich, A. Larsen, J. Marakis, G. Fytas, M. Klapper, K. Müllen, Controlled hydrogel fiber formation: The unique case of hexaphenylbenzene-poly(ethylene glycol) amphiphiles. *Small*, **10**(10), 1914–1919 (2014).

9. A. P. G. de Gennes, *Scaling Concepts in Polymer Physics*, Cornell University Press, (1979), p. 324.

10. I. A. Nyrkova, A. N. Semenov, J.-F. Joanny, A.R. Khokhlov, Highly anisotropic rigidity of "ribbon-like" polymers: I. Chain conformation in dilute solutions. *J. Phys. II*, (Paris), **6**, 1411–1428 (1996).

11. I. A. Nyrkova, A. N. Semenov, J.-F. Joanny, Highly anisotropic rigidity of "ribbon-like" polymers: II. Nematic phases in systems between two and three dimensions. *J. Phys. II*, (Paris), **7**, 825–846 (1997).

12. I. A. Nyrkova, A. N. Semenov, J.-F. Joanny, Highly anisotropic rigidity of "ribbon-like" polymers: III. Phase diagrams for solutions. *J. Phys. II*, (Paris), **7**, 847–875 (1997).

13. M. Rubinstein, R. H. Colby, *Polymer Physics*, Oxford University Press, NY, 2003.

14. T. Odijk, Stiff chains and filaments under tension. *Macromolecules*, **28**(20), 7016–7018 (1995).

15. H. Yamakawa, M. Fujii, Translational friction coefficient of wormlike chains. *Macromolecules*, **6**(3), 407–415 (1973).

16. Dynamic Light Scattering, Ed. W. Brown, Clarendon Press, Oxford, 1993, P. Russo, p. 523.

17. L. Harnau, R. G. Winkler, P. Reineker, Influence of polydispersity on the dynamic structure factor of macromolecules in dilute solution. *Macromolecules*, **32**(18), 5956–5960 (1999).

18. C. R. Rigler, E. S. Elson, *Fluorescence Correlation Spectroscopy: Theory and Applications*, Springer-Verlag: New York, 2001.

19. A. Brulet, F. Boue, J. P. Cotton, About the experimental determination of the persistence length of wormlike chains of polystyrene. *J. Phys. II*, **6**, 885–891 (1996).

20. G. Petekidis, D. Vlassopoulos, G. Fytas, R. Rülkens, G. Wegner, Orientation dynamics and correlations in hairy-rod polymers: Concentrated regime. *Macromolecules*, **31**(18), 6129–6138 (1998).

21. K. Harre, G. Wegner, Solution properties and kinetics of aggregation of an alkyl-substituted poly-(p-phenylene). *Polymer*, **47**, 7312–7317 (2006).

22. F. E. Golling, A. H. R. Koch, G. Fytas, K. Müllen, Synthesis and conformation of 3,6-connected cyclohexadiene chains. *Macromol. Rapid Commun.*, **36**(10), 898–902 (2015).

23. U. Scherf, K. Müllen, The synthesis of ladder polymers. *Adv. Polym. Sci.*, **123**(Synthesis and Photosynthesis), 1–40 (1995).

24. G. Petekidis, G. Fytas, U. Scherf, K. Müllen, G. Fleischer, Dynamics of poly(p-phenylene) ladder polymers in solution. *J. Polym. Sci. Polym. Phys.*, **37**(16), 2211–2220 (1999).

25. E. Somma, B. Loppinet, G. Fytas, S. Setayesh, J. Jacob, A. C. Grimsdale, K. Müllen, Collective orientation dynamics in semi-rigid polymers. *Colloid Polym. Sci.*, **282**, 867–873 (2004).

26. G. Fytas, H.G. Nothofer, U. Scherf, D. Vlassopoulos, G. Meier, Structure and dynamics of nondilute polyfluorene solutions. *Macromolecules*, **35**(2), 481–488 (2002).

27. M. Grell, D. D. C. Bradley, M. Chamberlain, M. Inbasekarn, E. WP. Woo, M. Soliman, Chain geometry, solution aggregation, and enhanced dichroism in the liquid-crystalline conjugated polymer poly(9,9-dioctylfluorene). *Acta Polym.*, **49**(8), 439–444 (1998).

28. T. Fütterer, T. Hellweg, G. H. Findenegg, J. Frahn, A. D. Schlüter, C. Böttcher, Self-assembly of amphiphilic poly(paraphenylene)s: Thermotropic phases, solution behavior, and monolayer films. *Langmuir*, **19**(16), 6537–6544 (2003).

29. A. M. Rosales, H. K. Murnen, S. R. Kline, R. N. Zuckermann, R. A. Segalman, Determination of the persistence length of helical and non-helical polypeptoids in solution. *Soft Matter*, **8**, 3673–3680 (2012).

30. M. Beer, M. Schmidt, M. Muthukumar, The electrostatic expansion of linear polyelectrolytes: Effects of gegenions, co-ions, and hydrophobicity. *Macromolecules*, **30**(26), 8375–8385 (1997).

31. H. Matsuoka, Y. Ogura, H. Yamaoka, Effect of counterion species on the dynamics of polystyrene sulfonate in aqueous solution as studied by dynamic light scattering. *J. Chem. Phys.*, **109**(14), 6125–6132 (1998).

32. Y. D. Zaroslov, G. Fytas, M. Pitsikalis, N. Hadjichristidis, O. E. Philippova, A. R. Khokhlov, Clusters of optimum size formed by hydrophobically associating polyelectrolyte in homogeneous solutions and In supernatant phase In equilibrium with macroscopic physical gel. *Macromol. Chem. Phys.*, **206**(1), 173–179 (2005).

33. S. Vanhee, R. Rulkens, U. Lehmann, C. Rosenauer, M. Schulze, W. Köhler, G. Wegner, Synthesis and characterization of rigid rod poly(*p*-phenylenes). *Macromolecules*, **29**(15), 5136–5142 (1996).

34. M. Bockstaller, R. Rülkens, W. Köhler, G. Wegner, D. Vlassopoulos, G. Fytas, Hierarchical structures of a synthetic rodlike polyelectrolyte in water. *Macromolecules*, **33**(11), 3951–3953 (2000).

35. M. Bockstaller, W. Köhler, G. Wegner, G. Fytas, Characterization of association colloids of amphiphilic poly(para-phenylene) sulfonates in aqueous solutions. *Macromolecules*, **34**(18), 6353–6358 (2001).

36. M. Bockstaller, W. Köhler, G. Wegner, D. Vlassopoulos, G. Fytas, Levels of structure formation in aqueous solutions of anisotropic association colloids consisting of rodlike polyelectrolytes. *Macromolecules*, **34**(18), 6359–6366 (2001).

37. A. Kröger, J. Belack, A. Larsen, G. Fytas, G. Wegner, Supramolecular structures in aqueous solutions of rigid polyelectrolytes with monovalent and divalent counterions. *Macromolecules*, **39**(20), 7098–7106 (2006).

38. A. Kröger, V. Deimende, J. Belack, I. Lieberwirth, G. Fytas, G. Wegner, Equilibrium length and shape of rodlike polyelectrolyte micelles in dilute aqueous solutions. *Macromolecules*, **40**(1), 105–115 (2007).

39. S. Wang, G. C. Bazan, Solvent-dependent aggregation of a water-soluble poly(fluorene) controls energy transfer to chromophore-labeled DNA. *Chem. Commun.*, **21**, 2508–2509 (2004).

40. B. W. Burchard, Solution properties of branched macromolecules. *Adv. Polym. Sci.*, **143** (*Branched Polymers* II), 113–194 (1999).

41. A. Narita, X. Feng, Y. Hernandez, S. A. Jensen, M. Bonn, H. Yang, I. A. Verzhbitskiy, C. Casiraghi, M. R. Hansen, A. H. R. Koch, G. Fytas, O. Ivasenko, B. Li, K. S. Mali, T.

Balandina, S. Mahesh, S. De Feyter, K. Müllen, Synthesis of structurally well-defined and liquid phase-processable graphene nanoribbons. *Nature Chem.*, 6, 126–132 (2014).

42. G. H. Fredrickson, Surfactant-induced lyotropic behavior of flexible polymer solutions. *Macromolecules*, 26(11), 2825–2831 (1993).

43. A. D. Schlüter, Dendrimers with polymeric core: Towards nanocylinders. *Top. Curr. Chem.*, 197, 165–191 (1998).

44. A. D. Schlüter, J. P. Rabe, Dendronized polymers: Synthesis, characterization, assembly at interfaces and manipulation. *Angew. Chem. Int. Ed.*, 39, 864–883 (2000).

45. S. Setayesh, A. C. Grimsdale, T. Weil, V. Enkelmann, K. Müllen, F. Meghdadi, E. J. W. List, G. Leising, Polyfluorenes with polyphenylene dendron side chains: Toward non-aggregating, light-emitting polymers. *J. Am. Chem. Soc.*, 123, 946–953 (2001).

46. F. Dutertre, K.-T. Bang, B. Loppinet, I. Choi, T. L. Choi, G. Fytas, Structure and dynamics of dendronized polymer solutions: Gaussian coil or macromolecular rod? *Macromolecules*, 49(7), 2731–2740 (2016).

47. C. Gstrein, B. Zhang, M. A. Abdel-Rahman, O. Bertran, C. Aleman, G. Wegner, A. D. Schlüter, Solvatochromism of dye-labeled dendronized polymers of generation numbers 1–4: Comparison to dendrimers. *Chem. Sciences*, 7(7), 4644–4652 (2016).

48. Z. Cao, J. M. Y. Carrillo, S. S. Sheiko, A. V. Dobrynin, Computer simulations of bottle brushes: From melts to soft networks. *Macromolecules*, 48, 5006–5015 (2015).

49. I. V. Mikhailov, A. A. Darinskii, E. B. Zhulina, O. V. Borisov, F. A. M. Leermakers, Persistence length of dendronized polymers: The self-consistent field theory. *Soft Matter*, 11, 9367–9378 (2015).

50. B. Loppinet, G. Fytas, G. Petekidis, T. Sato, G. Wegner, On the origin of the heterogeneous orientation dynamics of semiflexible polymers. *Europhys. J. E.*, 8(5), 461–464 (2002).

51. H. Gu, Y. Nakamura, T. Sato, A. Teramoto, M. M. Green, C. Andreola, Global conformations of chiral polyisocyanates in dilute solution. *Polymer*, 40(4), 849–856 (1999).

52. P. G. de Gennes, *The Physics of Liquid Crystals*, Oxford University Press, 1974.

53. P. G. de Gennes, Short-range order effects in the isotropic phase of nematics and cholesterics. *Mol. Cryst.*, 12(3), 193–214 (1971).

54. B. J. Berne, P. Pecora, *Dynamic Light Scattering*, Wiley NY (1976).

55. B. L. O'Steen, C. H. Wang, G. Fytas, Rayleigh-Brillouin scattering studies of the rotation-translation coupling and bulk viscosity relaxation of liquids composed of anisotropic molecules: p-anisaladehyde and aniline. *J. Chem. Phys.*, 80(8), 3774–3780 (1984).

56. C. Dreyfous, A. Aouadi, R. M. Pick, T. Berger, A. Patkowski, W. Steffen. Light scattering by transverse waves in supercooled liquids and application to meta-toluidine. *Eur. Phys. B*, 9, 401–419 (1999).

57. T. Sauer, G. Wegner. Small-angle X-ray scattering from dilute solutions of substituted phthalocyaninato-polysiloxanes. *Macromolecules*, 24, 2240–2252 (1991).

58. B. H. Zimm, J. K. Bragg, Theory of the phase transition between helix and random coil in polypeptide chains. *J. Chem. Phys.*, 31, 526–535 (1959).

59. D. L. Albertazzi, D. van der Zwaag, C. M. A. Leenders, R. Fitzner, R. W. van der Hofstad, E. W. Meijer, Probing exchange pathways in one-dimensional aggregates with super-resolution microscopy. *Science*, 344(6183), 491–495 (2014).

60. A. Vagias, L. Albertazzi, G. Fytas, E. W. Mejier unpublished.

61. S. F. Schulz, E. E. Maier, R. Krause, R. Weber, Dynamic light scattering on liquid-like polyelectrolyte solutions: correlation spectroscopy on dilute solutions of virus particles at very low ionic strength. *Prog. Colloid Polym. Sci.*, **81**, (Trends Colloid Interface Sci. **4**), 76–80 (1990).

62. S. Förster, M. Schmidt, Polyelectrolytes in solution. *Adv. Polym. Sci.*, **120**, 53–133 (1995).

63. S. D. Hudson, H.-T. Jung, V. Percec, W.-D. Cho, G. Johansson, G. Ungar, V. S. K. Balagurusamy, direct visualization of individual cylindrical and spherical supramolecular dendrimers. *Science*, **278**, 449–452 (1997).

64. V. Percec, J. Heck, G. Johansson, D. Tomazos, Towards tobacco mosaic virus-like self-assembled supramolecular architectures. *Macromol. Symp.*, **77**(1), 237–265 (1994).

65. A. Aggeli, G. Fytas, D. Vlassopoulos, T. C. B. McLeish, P. Mewer, N. Boden, Structure and dynamics of self assembling β-sheet peptide tapes by dynamic light scattering. *Biomacromolecules*, **2**(2), 378–388 (2001).

66. Y. Mai, A. Eisenberg, Self-assembly of block copolymers. *Chem. Soc. Rev.*, **41**(18), 5969–5985 (2012).

67. J. N. Israelachvili *Intermolecular and Surface Forces* (1991), p. 291.

68. D. E. Discher, A. Eisenberg, Materials science: Soft surfaces. Polymer vesicles. *Science*, **297**(5583), 967–973 (2002).

69. N. C. Forero-Martinez, B. Baumeier, K. Kremer, Persistence length dependence with backbone chemical composition and monomer sequence in phenylene polymers. To be published.

70. F. Schlütter, T. Nishiuchi, V. Enkelmann, K. Müllen, π-Congested poly(paraphenylene) from 2,2′,6,6′-tetraphenyl-1,1′biphenyl units: synthesis and structural characterization. *Polym. Chem.*, **4**(10), 2963–2967 (2013).

71. R. E. Kieltyka, A. C. H. Pape, L. Albertazzi, Y. Nakano, M. M. C. Bastings, I. K. Voets, P. Y. W. Dankers, E. W. Meijer, Mesoscale modulation supramolecular ureidopyrimidinone-based poly(ethylene glycol) transient networks in water. *J. Am. Chem. Soc.*, **135**(30), 11159–11164 (2013).

72. J. Marakis, K. Wunderlich, M. Klapper, D. Vlassopoulos, G. Fytas, K. Müllen, Strong physical hydrogels from fibrillar supramolecular assemblies of poly(ethylene glycol) functionalized hexaphenylbenzenes. *Macromolecules*, **49**(9), 3516–3525 (2016).

Chapter 2

Structure in Thin Films of π-Conjugated Semi-Conductors from the Perspective of Transmission Electron Microscopy

Martin Brinkmann

Institut Charles Sadron, CNRS — Université de Strasbourg,
23 rue du loess, 67034 Strasbourg Cedex, France

This chapter focuses on recent advances in the field of transmission electron microscopy (TEM) applied to the structural and morphological characterization of conjugated molecular and polymeric semi-conductors in thin films. Representative results are given for some key conjugated semi-conductors to illustrate the use of classical TEM observation modes (bright and dark field imaging, low dose high-resolution TEM, electron diffraction) as well as more advanced methods of investigations (electron tomography, cryo-TEM and energy filtered TEM) to unravel structurally challenging issues in plastic electronics.

1. Introduction

Thin films of π-conjugated molecular and polymeric semi-conductors are central in the field of plastic and molecular electronics. The success of these materials lies in (i) the possibility to fine-tune their electronic properties to a large extent via proper molecular and macromolecular engineering and (ii) their potential for large-area deposition on flexible substrates at a reduced cost.[1] π-conjugated semi-conductors are used as the active layers in numerous types of devices such as organic field effect transistors (OFETs), organic light emitting diodes, (OLEDs), organic solar cells (OSCs), photodetectors, sensors and more recently non volatile memory elements.[1-6] Prior to integrating new materials in devices, it is mandatory to draw proper correlations between their structural/morphological characteristics and their optical, electronic and opto-electronic properties.[7-14] This is because conjugated organic systems arrange in a large variety of complex structures and phases depending on the preparation conditions. For this reason, the processing-structure-property *nexus* is of central importance in plastic electronics.

Accordingly, the understanding of the structure and morphology of these materials, especially in thin films, is a necessary condition towards controlled and possibly improved device performances. Key structural parameters such as molecular packing, orientation of π-stacking directions in a film relative to metallic electrodes, average dimensions of crystals and their preferred contact plane on a substrate must be precisely controlled through adapted processing to optimize for instance charge transport properties in OFETs, OLEDs and OSCs. In OSCs devices based on the bulk heterojunction concept,[15,16] the formation of percolating networks, the mixing of phases and phase separation between electron-donor and electron-acceptor materials are essential phenomena to be visualized and ultimately controlled through adapted processing.[10]

Characterization of structure is required at different length scales from the molecular to the meso-scale. The structural investigation methods used to address these issues are numerous and based on the analysis of information in reciprocal space or in direct space, but seldom both simultaneously. To cite but a few: grazing incidence X-ray diffraction (GIXD),[17] numerous variants of scanning probe microscopy (atomic force microscopy, scanning tunneling microscopy, Kelvin probe microscopy, conducting AFM, etc...),[18,19] Scanning electron and transmission electron microscopies (SEM, TEM).

Non-local investigation methods such as grazing incidence X-ray diffraction are highly popular and effective to provide some of the important structural informations listed above but often require the use of synchrotron radiation, especially for semiconducting polymers, since their crystallinity and scattering intensity are low. One of the advantages of GIXD lies in the possibility to probe the structure of the organic layers at different depths in the films by controling the incidence angle of the X-ray beam.[12] This is of particular interest to evidence differences in structure and orientation of interfacial layers.[13,20,21] Recent developments in X-ray based scattering methods include X-ray microscopy which combines the scattering method with a spatially resolved rastering of a sample with a focused X-ray beam.[22,23]

Among all investigation techniques of thin film structure and morphology, TEM is possibly the most original method to unravel structural aspects in organic thin films. It allows for the combined analysis of structural features in reciprocal space via electron diffraction and morphological aspects in real space by using different 2D imaging modes, e.g., bright field and high resolution

TEM.[24-28] Moreover, TEM can probe submicrometric features with specific diffraction patterns, i.e., perform a local probe of structure in relation to morphology. In addition to these well established observation modes, new TEM operating modes were introduced in the last few decades by combining additional spectroscopic informations to morphology and structure. Energy filtered TEM is able to distinguish materials that differ in elemental composition, even in systems that are mainly carbon-based.[29,30] Electron tomography has become an indispensable tool to analyze the bulk 3D morphology in active layers such as bulk heterojunctions in OSCs.[31]

It is the purpose of this contribution to make a short, non exhaustive, survey of some representative TEM methods used in the last few decades to address important structural challenges on relevant π-conjugated materials used in plastic electronics. For each method, a few relevant and recent illustrations are provided.

2. Fundamental Aspects of Transmission Electron Microscopy

Transmission electron microscopy was developped originally by Knoll and Ruska in Berlin in 1931.[32] The principle of a transmission electron microscope is very similar to that of a corresponding light microscope, the major difference being that electrons are used instead of visible light, optical lenses are replaced by electromagnetic lenses and the image is visualized on a phosphorescent screen or a CCD (charge coupled device) camera. This microscopy is based on the strong interaction between electrons and matter which limits the maximum thickness of the samples to be analyzed by TEM to 100–200 nm. This thickness limitation imposes specific preparation methods for bulk samples and thin films. In the case of thin films, TEM analysis imposes to detach the semiconducting film from the substrate using appropriate preparation methods.[26,28]

The contrast in a TEM Bright field image is related to various sources of scattering of the incident electron beam off from the optical axis (these electrons are blocked by some diaphragms). Different sources of scattering can contribute to the contrast in a TEM bright field image.[33] Generally, electrons interact more with heavy than with light elements. This can be one limitation of using TEM on pure organic samples composed primarily of light elements (C, H, O, N). The first origin of contrast in TEM is therefore the so-called mass contrast: elements with a high Z give rise to a stronger contrast in BF. A second source of contrast is due to differences in thickness in the films: thicker parts of a sample appear darker in a BF image. Finally, diffraction contrast may further affect a BF image. This contrast is due either to the cutting-off of some of the diffracted electrons by an objective apperture positioned in the back focal pane or to the anomalous diffraction by crystal defects.[28] The intensity of the Bragg contrast depends strongly on the relative orientation of the incoming beam with respect to the crystal in the sample. Therefore, if a crystalline domain is highly diffracting the incoming electrons, and if the diffracted electrons are cut-off by an aperture, it can appear with a strong contrast compared to an amorphous region for instance. Most often, all these sources of scattering occur simultaneously, which makes the interpretation of contrast in BF images difficult. Additional observation modes may be used based on electron energy-loss spectroscopy (EELS)[29,30] or Scanning Transmission electron microscopy in high-angular annular dark field (STEM-HAADF).[34,35]

Due to the strong cross section interaction between the electrons in the TEM and organic matter, thin films are subjected to strong electron beam damage because of inelastic scattering interactions. The consequences of electron radiation are essentially chemical in nature, i.e., cross-linking of the carbon chains or hydrogen abstraction.[33] Therefore, organic semiconductors must be imaged under low irradiation and using most adapted electron energy.[34,35] The saturation dose for radiation damage D of organic compounds depends much on the chemical nature of the investigated molecules. For polynuclear aromatics and conjugated molecules, D is in the range 0.1–1 C/cm^2 whereas for aliphatics it is substantially lower, e.g., 10^{-2}–10^{-3} C/m^2 (100 kV beam, sample at 25°C).[28,36–38] This sets a strict limit to the observation conditions of molecular and polymer semiconductors bearing alkyl side chains such as regioregular poly(3-alkylthiophene). In addition, the insulating character of organic thin films can cause charging of the samples in the electron beam which imposes to coat the samples with a few nanometers of a thin amorphous carbon film. Noteworthy, concerns about radiation damage of organic semiconductors are less important for electron diffraction than for high resolution imaging.

3.   Classical TEM Observation Modes

3.1.   *Bright field and dark field TEM imaging*

As a first example, one may consider the case of regioregular poly(3-hexylthiophene) (P3HT), the working horse among all SCPs.[39] This polymer is made of a conjugated backbone of polythiophene bearing 3-hexyl side chains in a regioregular manner on each thiophene unit. P3HT is of particular interest as it is a semi-conducting polymer showing large hole mobilities (up to 0.1 $cm^2/V{\cdot}s$) and a characteristic semi-crystalline morphology consisting

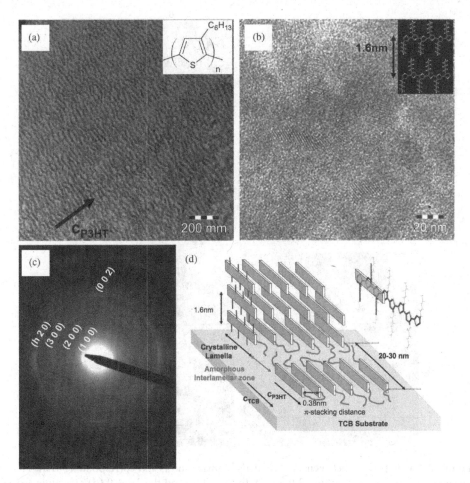

Figure 1: BF TEM (a), high-resolution–TEM (b), electron diffraction pattern (c) and schematic illustration of the semi-crystalline morphology (d) of regioregular poly(3-hexylthiophene) (P3HT). The inset in (b) represents the packing of two chains of P3HT separated by layers of 3-hexyl side chains. (Reproduced and adapted with permission from Ref. 42.)

of alternating crystalline domains and amorphous interlamellar zones (see Fig. 1).[40] To analyze properly the structure and semi-crystalline morphology of such polymers, it is possible to prepare highly oriented films using directional epitaxial crystallization.[40–43] The large areas of oriented P3HT films are perfectly suited for observations by TEM. The BF image in Fig. 1(a) shows the characteristic periodic lamellar structure of P3HT with the crystalline domains appearing as dark domains. Such a morphology is very similar to that observed for instance in epitaxied films of a polyolefin such as polyethylene.[44] The contrast in the BF image is due for one part to the electronic density difference

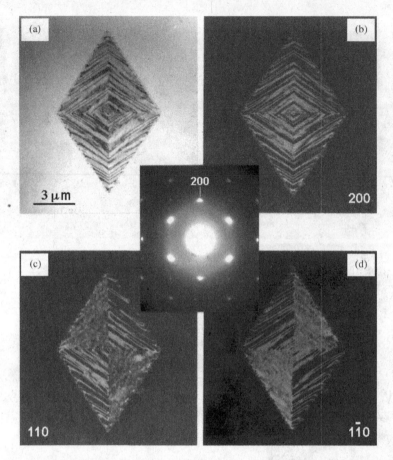

Figure 2: Comparison between the bright Field image in (a) and corresponding Dark Field images obtained by selecting the (2 0 0) (b), the 1 1 0 (c) and the 1 −1 0 (d) reflections from the electron diffraction pattern shown in the center of the image. (Reprinted with permission from Ref. 45 © 1993, V.C.H.)

between the crystalline and amorphous areas and for another part to a Bragg contrast because the crystalline domains diffract electrons strongly.

In BF TEM, the image is mainly formed from the electrons that pass close to the optical axis. In dark field (DF) imaging, the direct beam is blocked via an objective aperture and the aperture is centered on a given reflection in the diffraction pattern (see Fig. 2). In the DF image, those features responsible for the selected reflection are bright and the non diffracting domains are dark. Since the image is made with diffracted electrons, the DF image contains valuable information on orientation, defects and crystal dimensions. As a matter of fact, bright field and dark field images are complementary.

A remarkable example illustrating the strength in structural analysis of DF TEM in polymer science concerns the evidence of the sectorization of lozenge-shaped polyethylene single crystals using DF TEM by Lotz *et al.*[45] Figure 2 shows different views of a PE single crystal a few micrometers in size. Figure 2(a) features the characteristic lozenge-shape of a solution-grown PE crystal as observed in BF. It shows four sectors corresponding to parts of the crystals with different orientations of chain folds. Moreover, in each sector, dark stripes are seen. They result from Bragg contrast since the single crystals show a corrugated surface with alternating fold surface dip. This implies that stripes with different brightnesses in one sector of the crystal do not share exactly the same orientation with respect to the electron beam. Note how the DF pictures complement the BF image. This example illustrates nicely the capabilities of DF imaging applied to semi-crystalline polymers to evidence peculiar substructures in so-called "single-crystals".

In plastic electronics, DF proved very useful in better understanding the impact of molecular weight on the crystal habit of some poly(dialkyl)fluorenes such as poly(9,9'-dioctyl-fluorene- (PFO) or poly(9,9'-diethylhexylfluorene) (PF2/6) that can be used for their electroluminescence properties (see Fig. 3).[46–48] Contrary to P3HT that crystallizes with folded chains, polyfluorenes have a more rigid backbone. Its conformation changes from *all-trans* when the alkyl side chain is a linear *n*-octyl to helical for a branched 2-ethylhexyl side chain (see the inset in Fig. 3(b)). Such rigid chain polymers are expected to crystallize with extended chains in the form of lamellar crystals, the thickness of which should scale with the average molecular weight of the polymer. DF imaging was particularly well adapted to verify this assumption in the case of highly oriented polyfluorene films prepared on friction-transferred poly(tetrafluoroethylene) (PTFE). This is clearly demonstrated in Figure 3 that shows the electron diffraction pattern (a), the bright field image and two dark field images of two highly oriented PF26 films with different number average molecular weights (M_n). The DF images were obtained by selecting the very intense 0 0 21 reflection that corresponds to the repeat period of the monomer along the chain direction.[47] They show typical bright areas of the crystalline lamellae of PF2/6. For a given lamella, the bright domains seem to be discontinuous which simply indicates that some parts of the lamellae scatter less presumably because of some orientational distribution in the contact plane of the crystals. Moreover, the DF image shows that successive lamellae are separated by very faint dark lines corresponding to sharp grain boundaries. This is a consequence of the extended-chain mode of crystallization of these polyfluorenes for which no extended amorphous interlamellar zones are present

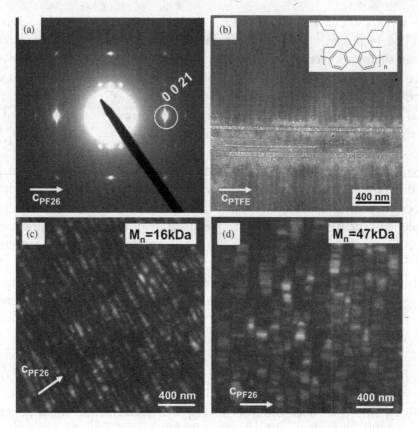

Figure 3: Structure of highly oriented films of PF26 prepared by epitaxial orientation of substrates of friction-transferred PTFE. (a) Diffraction pattern highlighting the 0 0 21 reflection used for the DF imaging. (b) BF showing lamellar crystals of PF26. (c) and (d): DF images of oriented PF26 films prepared from polymers with different molecular weights, 16 kDa (c) and 47 kDa (d). Note the strong increase in the average lamellar thickness with M_n. (Reproduced with permission from the American Chemical Society from Ref. 47).

as observed in P3HT. In addition, the statistical analysis of the average width of the PF2/6 lamellae indicates that it scales with the molecular weight of the polymer samples.[47]

3.2. *Low dose high resolution TEM*

The use of high resolution on organic molecular materials has been pioneered by Uyeda *et al.* and by Fryer *et al.* They obtained the first images with a direct visualization of the molecular packing at a molecular resolution in phthalocyanine thin films.[49,50] When the thickness of the films is sufficiently low (a few tens of nanometers), the electron beam interfers with itself when passing through the sample. The potential modulation caused by a regular arrangement of atoms or molecules in the sample "translates" to a corresponding phase shift of the scattered electrons. The phase modulation of the exiting beam cannot be directly detected — only the resulting modulation of the amplitude of the exiting beam is measured. Under such conditions, imaging in high resolution mode can provide molecular- or atomic-scale details but the contrast of the HRTEM image depends critically on the defocus which is used upon imaging.[28,51] HRTEM of organic materials is difficult since the lifetime of the materials under the beam is much reduced as compared to inorganic systems. Therefore, so-called low dose imaging is necessary to preserve, to the extent possible, the investigated area from potential beam damages. In the low dose operating mode, the incident electron beam is first deflected to an area near the one that will be imaged, in order to fine-tune all the proper settings and alignments. The beam is then redirected to the area of interest to acquire an image on a non-exposed area that should therefore be representative of the pristine film.

This technique is suited to observe local structural defects in molecular crystals such as pentacene or phthalocyanines.[52,53] Structural defects such as edge and screw dislocations, lattice curvature as well as grain boundaries are detrimental for charge transport in thin film devices.[54,55] HRTEM is a mean to observe such stacking defects in thin films of semiconductor crystals such as pentacene. Pentacene is a molecular semiconductor of the family of acenes and is formed by a succession of five fused benzene rings (see Fig. 4(a)). Pentacene has attracted much attention as a molecular semiconductor showing hole mobilities in the range 0.1–1 $cm^2/V \cdot s$.[56] In thin films, pentacene crystallizes in various polymorphs including a specific so-called thin film phase.[57] Molecules form a layered structure with a herringbone arrangement of molecules in the layers. Intermolecular interactions are strong in the layers but week between successive layers. Therefore, defects such as edge dislocations (Fig. 4(c) and (e)) or lattice curvature (Fig. 4(d)) are often observed in these pentacene thin films. Martin and coworkers observed characteristic grain boundaries in pentacene by HRTEM.[52] In Fig. 4(b), such a grain boundary is shown with a boundary tilt angle of 40° between the two crystalline subdomains. Moreover, HRTEM shows the effective width of the grain boundary at different locations (two arrows in Fig. 4(b)).

HRTEM is also very useful to better visualize the semi-crystalline morphology in thin films of several poly(alkylthiophene) (P3ATs) and its evolution with annealing temperature or other processing conditions.[58–60] As seen in Fig. 1, BF can already help distinguish crystalline and amorphous domains of

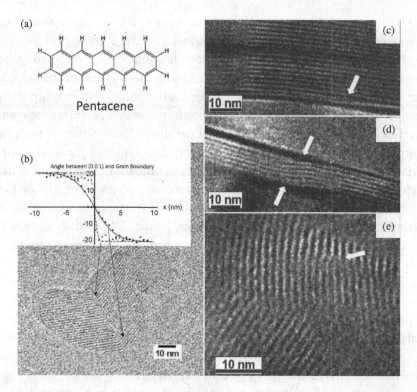

Figure 4: (a) Chemical formula of pentacene, (b) HR-TEM image of a high-tilt grain boundary in a pentacene polycrystal. The fringes correspond to 1.44 nm (0 0 1) planes. The plot shows the angle between the (0 0 1) normal and the grain boundary plane as a function of the distance x from the grain boundaries. (c), (d) and (e) show additional stacking defects in evaporated thin films of pentacene on oriented poly(tetrafluoroethylene) substrate. Adapted with permission from reference (52) (Wiley-VCH) and reference (53) (American Chemical society).

P3HT. However, HRTEM can also be used for that purpose when crystalline domains of P3ATs are oriented face-on the substrate (see Fig. 1(d)). In that case, the P3HT chains lie with their backbone flat-on on the substrate and successive chains are π-stacking along the normal to the film plane. In this case, when observed by TEM, the incident electron beam is oriented parallel to the π-stacking direction. A contrast in electron density exists between layers of π-stacked polythiophene backbones and layers of alkyl side chains. This gives rise in the HRTEM images to a fringe pattern with a periodicity of 1.6–1.7 nm corresponding to 1 0 0 planes. HRTEM distinguishes between crystalline domains with face-on orientation and domains with other orientations or with an amorphous structure that will not show a fringe pattern. More interestingly, HRTEM helps to measure the average size of crystalline domains along the chain direction, l_c, and also along the alkyl side chains, l_a (see Figs. 5

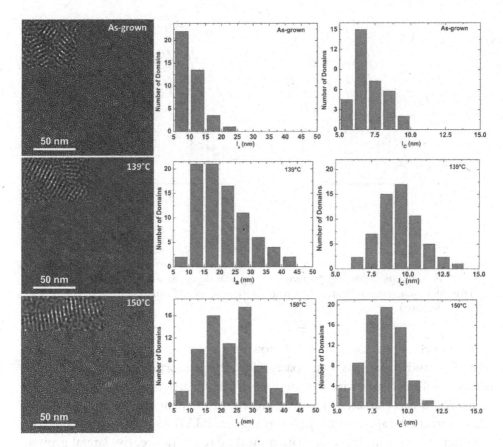

Figure 5: Evolution of the nanomorphology in P3OT thin films as a function of increasing annealing temperature (middle-isothermal) and corresponding histograms of the domain size l_a (lateral dimension along the a_{P3OT} axis) and l_c (stem length). (Reprinted with permission from Ref. 59 © 2012, American Chemical Society.)

and 6). Statistical analysis of HRTEM images as a function of preparation conditions (e.g. thermal annealing), can help understand the correlation between processing, domain coarsening and ultimately charge transport in the films. By contrast, GIXD studies on P3ATs cannot give any information on the level of order in the direction of the polythiophene backbone because no 0 0 l reflection is accessible in the GIXD patterns (incidentally, and in addition, the 0 2 0 and the 0 0 2 reflections of P3HT overlap in GIXD patterns of non oriented films). Therefore, HRTEM is a valuable alternative method to learn more about ordering phenomena along the chain direction, especially when highly oriented films are analyzed.

Figure 5 shows some characteristic HRTEM images of representative poly(3-octylthiophene) (P3OT) spin-coated thin films as a function of the

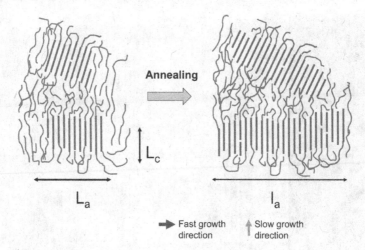

Figure 6: Schematic illustration of the anisotropic lamellar growth in P3ATs as inferred from HRTEM results. Annealing promotes fast growth along the side chain direction whereas lamellar thickening remains limited. (Reprinted with permission from Ref. 59 ©2012, American Chemical Society.)

temperature of annealing.[59] The following trends were identified from these HRTEM studies: (i) in as-spin-coated films, the proportion of face-on oriented nanocrystals increases with the number of carbon atoms in the side chain, (ii) annealing favors the lateral in-plane growth of the nanocrystals along the side chain direction (a_{P3AT} axis), (iii) for a given P3AT, the proportion of face-on oriented domains increases with annealing temperature, (iv) lateral growth along the a_{P3AT} axis is most efficient for the longer octyl side chains, and (v) thermal annealing induces only modest lamellar thickening. This thickening is limited by the poor sliding diffusion of π-stacked P3AT chains whereas for n-alkyl side chains, lateral growth is favored by weak van der Waals interactions. In other words, as illustrated in Fig. 6, the side chain direction a_{P3HT} corresponds to a fast growth direction of P3HT crystals during thermal annealing contrary to c_{P3HT}. Regarding charge transport in P3HT, it is most effective along the polythiophene backbone and π-stacking direction. The absence of lamellar thickening therefore explains at least in part the limited change in charge transport induced by thermal annealing.[59]

Beside fundamental aspects of growth and structure of molecular and macromoleculcular semiconductors, HRTEM has been used in the last decades for the structural analysis of the active layer structure in devices, especially in organic solar cells.[61] The concept of BHJ solar cells introduced by Heeger and coworkers was an incentive to use TEM and especially TEM tomography and HRTEM to characterize the nanomorphology in such solar

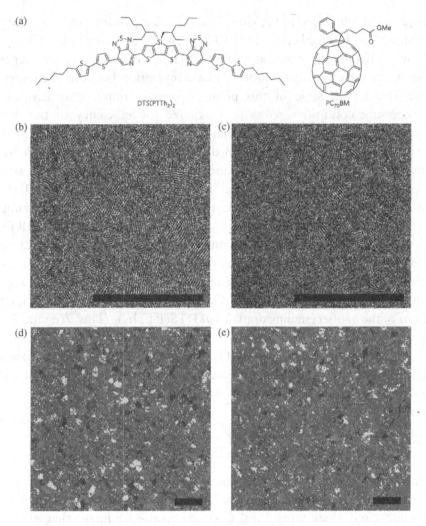

Figure 7: (a) Molecular structures of DTS(PTTh$_2$)$_2$ and PC70BM used for the elaboration of BHJ solar cells. (b) and (c) High-magnification TEM images showing lattice fringes from in-plane stacking of the DTS(PTTh$_2$)$_2$ phase: BHJ film cast from a pure solvent (b) and addition of 0.25% v/v DIO solvent (c). (d) and (e) Lower magnification, false colour images encoding the direction and size of crystalline regions for BHJ films cast from a pure solvent (d) and with addition of 0.25% v/v DIO solvent additive (e). The domain size is reduced when adding 0.25% v/v DIO in the solvent (optimal device) relative to the film cast from a pure solvent. Scale bars, 100 nm. (Adapted from Ref. 61 with permission from NPG, 2012.)

cells. Organic solar cells involve an active layer made of two materials with complementary electron-donor and electron–acceptor properties. Two representative molecular systems used to prepare high efficiency solar cells are shown in Fig. 7. They are a small-molecule donor: 5,5′-bis{(4-(7-hexylthiophen-2-yl)thiophen-2-yl)-[12,5]thiadiazolo[3,4-c]pyridine}-3,3′-di 2-ethy-lhexylsilene-2,2′-bithiophene DTS(PTTh$_2$)$_2$ and an acceptor : PC$_{70}$BM.

When active layers combining DTS(PTTh$_2$)$_2$ and PC$_{70}$BM are prepared by spin-coating from blend solutions, phase separation between the materials occurs. The length scale of this phase separation must be controlled by the processing conditions so as to maximize the efficiency of the device performances. It has been observed that the addition of small amounts (0.25% in weight) of additives such as 1,8-diiodooctane (DIO) to the blend of the materials in chlorobenzene enhances substantially the efficiency of the devices. HRTEM helps visualize the influence of the addition of DIO on the nanomorphology of the active layer. As seen in Fig. 7, the stacking of DTS(PTTh$_2$)$_2$ molecules results in well defined fringe patterns in the HRTEM images, with a typical periodicity of 2 nm. Such fringe patterns arise because of the electron density difference between the electron-rich core of DTS(PTTh$_2$)$_2$ and the lateral alkyl side chains. To visualize more readily the changes in domain size in the BHJ, processing of the HRTEM image was used to determine the contours of the nanocrystalline domains of DTS(PTTh$_2$)$_2$ (Figs. 7(d) and 7(e)). Such an analysis reveals the change in average domain size for DTS(PTTh$_2$)$_2$ as a function of the addition of DIO. Addition of DIO helps decrease the domain size of DTS(PTTh$_2$)$_2$ observed by HRTEM down to dimensions favorable for effective exciton dissociation and leading to improved device efficiency.[61]

3.3. *Electron diffraction*

Electron diffraction in a TEM is a powerful method to analyze the structure of organic thin films. As seen next, it helps determine preferred orientations in thin films and phase transitions. Contrary to the BF image that is formed in the image plane of the objective lens, the diffraction pattern is formed in the back-focal plane. The incident electron wave is scattered by the atomic potential of the sample to form the diffraction spots in the back focal plane. Electron diffraction can be performed on very small areas, down to a ∼100 nm in diameter in the so-called selected area electron diffraction (SAED) mode. However, the beam sensitivity of the molecular and polymeric semiconductors requires to work with low electron doses, especially for all organic semiconductors bearing solubilizing alkyl side chains. The SAED mode must be distinguished from the microdiffraction mode that uses a convergent conical electron beam ($\alpha > 10^{-3}$ rad) and has not yet been used extensively on organic and beam sensitive systems.

Preferred orientations in thin films

The epitaxial growth of certain organic semiconductors can result in the coexistence of different orientations of the crystals on the substrate. This is illustrated with two examples, namely (a) tris(8-hydroxyquinoline) aluminum or gallium (Alq3 and Gaq3, respectively)[62,63] oriented on PTFE and (b) P3HT oriented on potassium bromobenzoate substrate.[64]

Alq3 is a key material widely used in the elaboration of OLEDs. The evaporation of Alq3 and Gaq3 usually results in amorphous thin films. However, when substrates of oriented PTFE are used, Moulin and coworkers showed that the evaporation of Alq3 on oriented PTFE substrates heated at 100°C, generates elongated Alq3 crystals because of the unique nucleating ability of crystalline PTFE. The films are made of crystals with their long axis oriented parallel and at ±60° to the PTFE chain direction. This distribution of in-plane orientations is reflected in the ED pattern in Fig. 8(b). After analysis of the SAED pattern of a single Alq3 crystal, it was possible to interpret the ED pattern and propose a scheme illustrated in Fig. 8(c) to account for the three crystal orientations. The expected ED pattern for an Alq3 crystal with a (0 1 0) contact plane is shown in Fig. 8(d). Accordingly, the existence of three orientations of Alq3 crystals on PTFE was explained by a mechanism of 1D epitaxy. Alq3 molecules are spherical objects that arrange at a distance of 0.625 nm along the 1 0 0 direction whereas the surface of PTFE is described by an array of rods separated by 0.565 nm. If one molecule is placed in a groove on the PTFE substrate, there are two possibilities to place the next neighbour molecule in a favorable site: either in the same groove, i.e., parallel to c_{PTFE}, or in the next groove at ±60° to c_{PTFE} because $a_{PTFE}/a_{Alq3} \sim \sin(60°)$.

Beside Alq3 and Gaq3, numerous other examples of epitaxial growth and determination of contact planes as a function of the preparation conditions (substrate temperature, nature of the substrate e.g. mica and SiO_2, deposition rate and film thickness) can be found in the literature for molecular and polymeric materials.[63–74] They all illustrate the essential role of TEM and especially electron diffraction to provide new insight in the growth mechanisms of semi-conducting molecules and polymers on orienting substrates especially when TEM is combined with X-ray diffraction methods.

Epitaxied films of P3HT on substrates of potassium bromo-benzoate (KBrBz) provide another representative example illustrating the strength of electron diffraction to analyze the orientation mechanism of an SCP.[74] Epitaxy makes use of orienting substrates to induce preferential growth directions of a SCPs. SCPs such as P3ATs usually have rather high melting temperatures

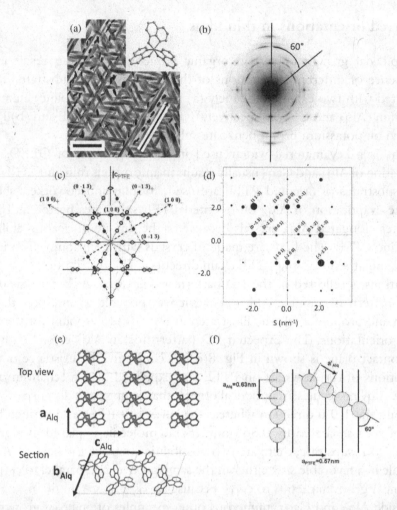

Figure 8: (a) BF TEM image of a highly oriented Alq$_3$ film grown on an oriented PTFE substrate. The scale bar corresponds to 1 μm. The inset shows the molecular structure of Alq$_3$. (b) Electron diffraction pattern. (c) Schematic illustration of the ED pattern. It consists of the overlap of three ED patterns with the 100 peak oriented parallel and at $\pm 60°$ to c_{PTFE}. (d) Calculated ED pattern for an Alq$_3$ crystal (alpha phase) with a (0 1 0) contact plane i.e. a [0 3 1] zone. (e) Top view and section view of the contact plane of Alq$_3$ at the interface with PTFE. (f) Illustration of the two orientations of aligned Alq$_3$ molecules (shown as yellow circles) in the parallel grooves of the PTFE substrate. (Adapted from Ref. 62 with permission from Wiley-VCH, 2002.)

(240°C for P3HT) which limits strongly the type of substrates that can be used for orientation. Interestingly, aromatic salt crystals such as potassium acid phthalate or potassium bromo-benzoate can be used as substrates for the orientation of SCPs because (i) they can be used at very high melting temperatures and (ii) their cleavage surface consists of a regular array of aromatic groups that is ideal for the epitaxial orientation of polymers.[74] In addition they are quite soluble and can be removed from the SCP films by rinsing in solvents such as ethanol.

As seen in Fig. 9(a), highly oriented and nanotextured P3HT thin films are formed after isothermal annealing at 180°C on KBrBz substrates. The P3HT films consist of a network of interconnected semi-crystalline domains oriented along two preferential in-plane directions as shown in Fig. 9(b). Isothermal crystallization of regioregular P3HT on potassium 4-bromobenzoate (K-BrBz) crystals generates extended domains of highly interconnected crystalline P3HT domains oriented along two preferential in-plane directions on the surface of KBrBz crystals. Analysis of this ED pattern helps establish the growth mechanism of this original nanotextured morphology. The ED pattern in Fig. 9(c) indicates that the crystals grow with a unique $(1\,0\,0)_{P3HT}$ contact plane on K-BrBz (only 0 k l reflections are observed). The schematic representation of the ED pattern in Figure 9(d) shows that two ED patterns overlap. They are oriented 70° apart, as observed for the two orientations of P3HT stripes in BF. This 70° angle can be explained by the fact that the nucleation of oriented P3HT occurs at step edges of the K-BrBz substrate i.e. along two preferential in-plane directions, namely $[0\,2\,1]_{K-BrBz}$ and $[0\,-2\,1]_{K-BrBz}$ which make a relative angle of 70°. This nucleation is further favored by the fact that the average height of the $(1\,0\,0)_{K-BrBz}$ substrate terraces matches the 1.6 nm layer period of π-stacked P3HT chains. Such nanotextured and highly crystalline P3HT films could be of interest for the elaboration of active layers with a nanostructured bulk heterojunction morphology used in organic solar cells.

Structure determination of SCP polymorphs using electron diffraction

"Electron crystallography" based on electron diffraction data obtained from polymer single crystals has been used to determine the structure of various mechanically unstable polymorphs of polyolefins.[38,75,76] It requires however relatively large single crystals (a few microns in size), easily obtained for e.g. the frustrated β phase of isotactic polypropylene.[38,76-78] A similar procedure is more problematic for SCPs that form typically smaller crystals providing insufficient diffraction data to determine a structure. So far, the only example of structure determination by electron diffraction on SCP single crystals concerns form II of P3HT, the crystals of which were prepared by self-seeding.[79]

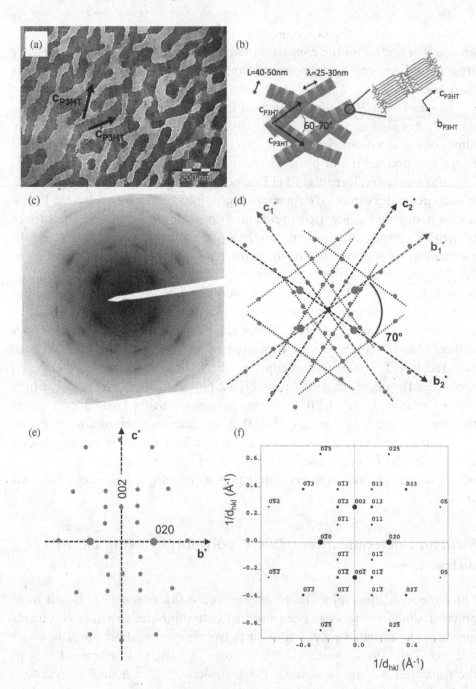

Figure 9:

Figure 9: (figure on facing page) (a) BF TEM image of oriented domains of P3HT prepared by isothermal crystallization at 180°C on crystalline substrates of KBrBz. (b) Schematic illustration of the semi-crystalline structure within stripes of P3HT where bright and dark stripes alternate and correspond to amorphous and crystalline domains respectively (crystalline in red and amorphous in blue). (c) experimental ED pattern of epitaxied P3HT film. (d) Interpretation of the experimental ED pattern by the sum of two ED patterns rotated by 70° and shown in red and green. The axes b_i^* and c_i^* with i = 1,2 designate the two orientations of the crystalline P3HT domains observed on KBrBz substrates. (e) representation of the characteristic ED pattern corresponding to the [1 0 0] zone. (f) calculated [1 0 0] zone for a P3HT crystal (form I). (Reprinted with permission from 74 © 2009, American Chemical Society.)

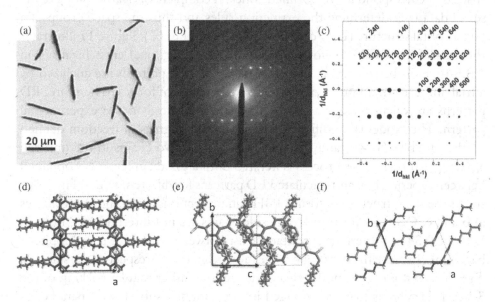

Figure 10: (a) POM image of single crystals of P3HT grown by self-seeding. (b) Selected area electron diffraction pattern of a form II P3HT single crystal. (c) calculated ED pattern for the refined crystal structure shown in (d),(e) and (f). (Adapted with permission from Ref. 79, Wiley-VCH, 2012.)

Figure 10 shows the electron diffraction pattern for a single crystal of form II — P3HT as well as the calculated ED pattern and the refined structure. Refinement of the structure was based on a trial-and-error approach using a tentative model proposed earlier for the structure of regioregular poly(3-butylthiophene) (P3BT) by Meille and coworkers.[80] Most importantly, this structural refinement allows to draw the major characteristics of the model in terms of π-stacking and interdigitation of 3-hexyl side chains.

Although, form II could be refined from single crystal data, such large well defined single crystals are so far the exception rather than the rule for conjugated polymers. An elegant alternative approach consists of generating large oriented domains of SCPs in highly oriented thin films as demonstrated by Brinkmann and coworkers.[40,41,46,81] Various methods to prepare highly oriented films of SCPs were tested e.g. directional epitaxial crystallization,[40,41,46] high-temperature rubbing [82–85] and on friction-transferred PTFE substrates.[47,86] Most interesting are SCP films in which all crystals share the same contact plane and a common in-plane orientation. Such films provide sharp or slightly arced reflections and can be studied by rotation-tilt in a TEM. This technique allows to record different projections in reciprocal space that correspond to well defined zones. A complete or nearly complete 3D set of data help determine the extinction rules and thus the space group of a given crystal structure ($P2_1/c$ in the case of form I P3HT). Figure 11 illustrates the case of form I P3HT for films oriented by directional crystallization on 1,3,5-trichlorobenzene.[40,41] The electron diffraction patterns are analyzed by a trial-and-error method to determine the structure of form I. Therefore, ED patterns are calculated for a given zone and compared to the experimental pattern. The model is modified following certain degrees of freedom defined by the variables $\theta_1 - \theta_4$ and the shift between the two chains in the unit cell δ_c (see Fig. 11(a)). This process is iterated until a good agreement is obtained between experimental and calculated ED patterns for different zones. Figure 11 shows the final structure of form I P3HT in projections along the unit cell axes **a**, **b** and **c**. Contrary to form II, the side chains are not interdigitated in form I. Short 0.34 nm interplanar contacts exist between successive polythiophene backbones because the backbone planes are inclined with respect to the stacking direction (**b** axis). This model also evidences the rather regular packing of the 3-hexyl side chains that are arranged in a rectangular subcell with parameters $a_s = 0.7$ nm and $b_s = 0.78$ nm.[81]

More recent structural investigations using TEM focus on polymers of increasing complexity. Indeed, the need for better matching between the absorption spectrum of the polymers and the solar spectrum led to the synthesis of so-called low bandgap polymers.[87] They are alternating copolymers poly(A-B) where A and B correspond to electron-rich and electron-deficient monomers. The proper choice of the two monomers allows to adjust precisely HOMO and LUMO levels as well as the resulting bandgap. Moreover, when rylene diimides (perylene bisimide or naphthalene bisimide) alternate with bithiophene, polymers such as p(NDI2OD-T2) can even present a characteristic n-type charge transport.[88,89] To establish structure-property correlations, it is important to determine how the two alternating blocks in the backbone are π-stacked in a crystal: extreme cases correspond to segregated and mixed stacking modes of A and B units.[86] Unravelling the structure of such polymers is a challenge given their chemical complexity and the limited level of crystallinity that is usually observed in thin films.

However, there exist means to bring such complex macromolecular systems to high order. Controlled growth methods such as solvent vapor annealing or epitaxial crytallization allow to grow submicrometric crystals that provide single-crystal like diffraction patterns suitable for structural refinement. Moreover, adapted alignment methods such as high-temperature rubbing are useful to grow oriented systems that can display fiber symmetry and hence help access the relative stacking of A and B monomers. Representative structures were recently proposed for p(NDI2OD-T2), PCPDTBT and F-PCPDTBT.[82,86] Electron diffraction clearly established the polymorphism of p(NDI2OD-T2) with two structures corresponding to segregated (form I) and mixed (form II) modes of stacking of NDI and bithiophene units as illustrated in Fig. 12. In the case of PCPDTBT and F-PCPDTBT, pairs of chains were found to stack in a segregated mode. Such dimer-like structures were found to arrange in a highly symmetric structure that did not show long-range π-stacking, similarly to the case of the α form of PFO.[46] The precise structural refinement of both structures including the determination of the space group were only possible by combining the structural information of single-crystal-like and oriented fiber-like patterns obtained by electron diffraction that are shown in Fig. 13. Notably, the fiber patterns were essential to discriminate between segregated and mixed type of stacking of the PCPDTBT and F-PCPDTBT chains. Indeed, the intensities of the reflections in the successive layer lines along the chain direction indicate the relative shift between chains in the unit cell.[86]

4. Advanced TEM Imaging Modes

4.1. *TEM tomography*

Classical BF TEM provides an image that corresponds to a 2D projection of a 3D structure. TEM tomography has been developped with the aim of providing a clearer insight into the 3D structure of the objects observed in a TEM.[90–93] To establish a 3D structure in a TEM, it is necessary to first record a sequence of 2D images obtained at different tilts (typically 100 images). To this aim, the sample is tilted around a single axis using either a constant angular step or using a so-called Saxton scheme.[93] Most importantly, upon tilting, all images must be carefully set to a common tilt axis and lens distortions must be corrected.

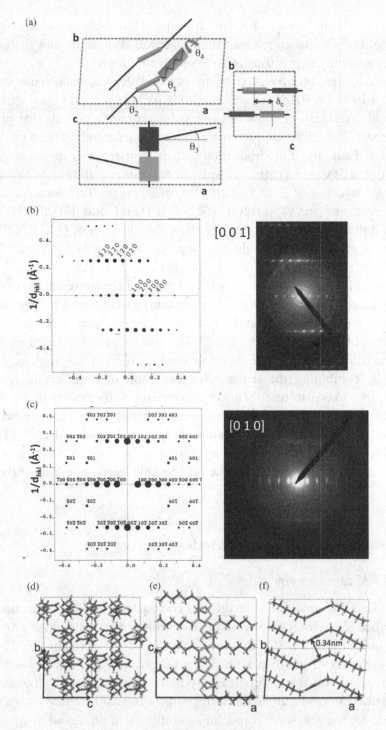

Figure 11:

Figure 11: (Figure on facing page) (a) Schematic illustration of the P3HT structural model defining the angles $\theta_1 - \theta_4$ used for the refinement. (b) and (c) calculated (left) and experimental (right) electron diffraction patterns corresponding to the determined crystal structure of form I P3HT for the [0 0 1] (b) and [0 1 0] (c) zone axes. (d), (e) and (f) Projections of the refined P3HT structure along the **a**, **b** and **c** axes of the unit cell. (Reprinted with permission from Ref. 81 © 2010, American Chemical Society.)

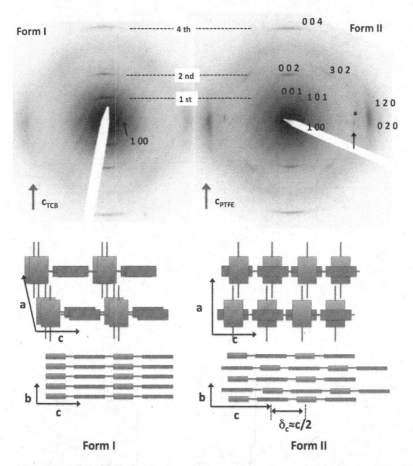

Figure 12: Comparsion of the electron diffraction patterns of p(NDI2OD-T2) films in form I (left) and form II (right) prepared by epitaxy on 1,3,5-trichlorobenzene and by melt-recrystallization on oriented poly(tetrafluoroethylene) substrates, respectively. Below, a schematic illustration shows the different stacking modes determined for the two polymorphs. (Reproduced with permission from Ref. 86 @ 2012, American Chemical Society.)

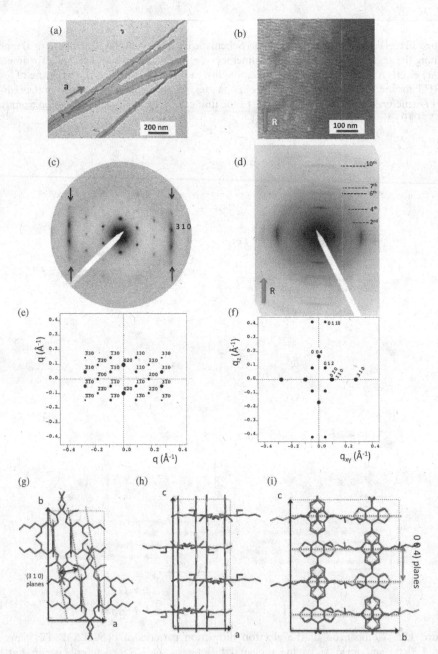

Figure 13:   Bright field images (a and b), experimental ED patterns (c,d) and calculated ED patterns (e, f) F- PCPDTBT samples prepared by solvent vapor annealing (a,c,e) and by high-temperature rubbing at 240°C (the rubbing direction R is indicated by an arrow in b and d) (b,d,f). The projections of the refined structure along the **c**, **b** and **a** axes of the unit cell are shown in g, h and i, respectively. (Reproduced with permission from Ref. 82, American Chemical Society, 2015.)

In a second step, a 3D reconstructed object is built out of the sequence of TEM 2D images. The most widely used method for the reconstruction is a so-called real-space back-projection technique.

Several examples in the recent literature illustrate the benefits of using electron tomography in plastic electronics, especially in organic photovoltaics. Uncovering the 3D morphology in the active layer of a BHJ solar cell is a key step in the possible improvement of organic solar cells. In a typical BHJ layer, both electron-donor and electron–acceptor materials are blended together and should form co-continuous networks with a nanoscale phase separation in order to allow (i) efficient exciton dissociation at a donor/acceptor interface and (ii) effective charge transport through percolating networks of donor and acceptor domains until charges are collected by the electrodes.[94]

Pioneering work in the field of TEM tomography on BHJ solar cells was made by Loos and coworkers on blends of MDMO-PPV/PCBM (MDMO-PPV: poly[2-methoxy-5-(3′,7′-dimethyloctyloxy)-1,4-phenylene vinylene]).[31,95–97] Figure 14 compares the classical BF TEM image (a) and the snapshot of a volume reconstruction obtained by electron tomography (b) of a MDMO-PPV/PCBM film (80 wt.-% PCBM). Classical BF TEM shows dark PCBM domains embedded in a polymer-rich phase but no evidence of connections between the PCBM domains. TEM tomography instead reveal the existence of PCBM-rich strands connecting the PCBM-rich domains, and thus provides the first evidence for a percolating nanoscale network. Such

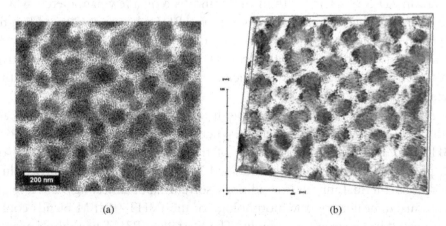

(a)   (b)

Figure 14: BF TEM image (a) and the snapshot of a volume reconstruction obtained by electron tomography (b) of a MDMO-PPV/PCBM film (80 wt.-% PCBM). The volume dimensions are 1130 nm × 970 nm × 80 nm. (Reprinted with permission from Ref. 96 @ RSC.)

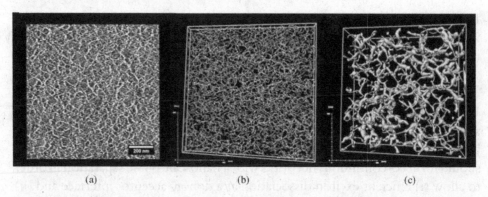

(a)   (b)   (c)

Figure 15:   Result of electron tomography applied to active layers of PF10TBT/PCBM. (a) one slice of the (x, y) — plane of the final 3D dataset with bright-looking polymer strands and darker PCBM-rich regions. (b) snapshot of the corresponding volume reconstruction showing the existence of a 3D nanoscale polymer network (volume dimensions are 1112 nm × 1090 nm × 75 nm), the polymer strands are made thinner than they are in reality to facilitate vizualisation. (c) zoom-in of the same volume (with volume dimensions of 227 nm × 227 nm × 75 nm and the thickness of the polymer strands to scale). (Reproduced with permission from Ref. 96.)

percolation is essential to ensure a good charge transport in the active layer towards the electrode of the device.[94]

In yet another example, Loos and coworkers demonstrated the existence of a 3D network of a polymer-rich phase (PF10TBT: poly[2,7-(9,9-didecyl-fluorene)-*alt*-5,5-(4,7-di-2-thienyl-2,1,3-benzothiadiazole)]) in a blend film with PCBM. PF10TBT forms a polymer-rich phase with highly interconnected strands in 3D.[7,95,96] The width of the strands is only a few nanometers which is essential for the exciton dissociation in the BHJ whereas the strand length can reach several tens of nanometers. Moreover, the polymer rich 3D nanoscale network is also highly effective to channel holes out of the device to the positive ITO electrode whereas the PCBM-rich regions can transport electrons to the metal electrode.

A last example showing the strength of the tomographic approach deals with P3HT-PCBM BHJ solar cells.[97] When BHJ solar cells are prepared from P3HT and PCBM, the morphology of the devices and the crystallinity of the phases can be substantially modified by the post-deposition processing conditions (thermal annealing conditions, solvent vapor annealing). The impact of thermal annealing on the morphology of the P3HT/PCBM blends could be observed by electron tomography. The crystalline P3HT nanofibrils grown in the thermally annealed films form a highly interconnected 3D network. Most importantly, ET demonstrates that the nanofibrils are inclined in the volume of the films, which is beneficial for the transport of holes through the entire thickness of the device. In addition, ET has revealed the existence of a composition gradient in the layers along the normal to the film surface. The quantity of nanofibrils could be quantified as a function of the z-position in the films. At the ITO electrode, an enrichment of P3HT is observed whereas a corresponding enrichment of PCBM is observed at the other metal interface of the device. Such a distribution of the two components in the bulk of the active layers is thought to be highly beneficial for charge separation, transport and extraction.

4.2. Cryo-TEM

Many SCPs such as P3HT are reported to form nanofibrils (or nanowires) in organic solvents as a consequence of aggregation. The nanofibrils are several micometers in length and have a characteristic diameter of the order of 10–30 nm. Such nanofibrils are highly crystalline and there exists various methods to precisely control their nucleation and growth in solution.[98,99] Nucleation can be controlled by the so-called self-seeding method whereby a partial melting of a dispersion of crystals in solution at a given temperature leaves some seeds that can subsequently grow to form large nanofibrils when the solution is cooled down. Growth can be controlled in solution by adjusting the nature of the solvent or the crystallization temperature. The precise control of nanofibrils in solution results in improved performances of photovoltaic devices, in particular, in higher power conversion efficiencies.[99] However, most structural studies were performed on the "dried" films of such nanofibrils after solvent evaporation. Given the possible inclusion of solvent molecules in the structure of nanofibrils, it is essential to characterize the structure of such samples in organic solvents prior to drying. Cryo-TEM is a technique that allows to perform structural studies on soft nano-objects formed in solution avoiding all possible structural alteration resulting from a drying process.[100] This technique is well known in aqueous media but less used in organic solvents. The principle of the technique implies a freezing of the solution containing the nano-objects so as to induce the vitrification of the solvent matrix and to avoid its crystallization. As a result, the nano-objects are left unaltered in the amorphous matrix of the solvent. This implies specific cooling conditions of the organic solvent.

Wirix and coworkers showed that cryo-TEM can be successfully applied to P3HT nanofibrils in toluene.[101,102] Figure 16 shows the characteristic TEM micrographs of vitrified solutions of P3HT (1% wt) after ageing for >7 days in toluene (a) and oDCB (b). The same samples are observed after dropcasting on carbon substrates in (c) and (d). The comparison of the TEM images obtained

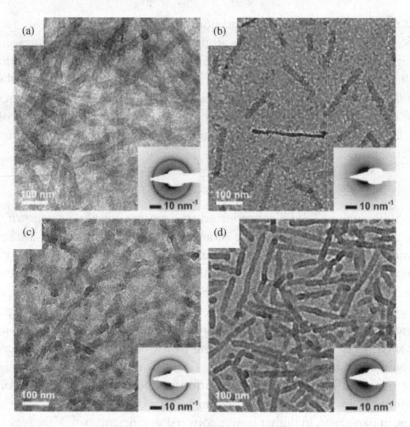

Figure 16: TEM micrographs of vitrified solutions of P3HT (1% wt) after ageing for >7 days in toluene (a) and oDCB (b). The same samples are observed after dropcasting on carbon substrates in (c) and (d). All insets correspond to the ED patterns. (Reproduced with permission from Ref. 101, © 2014, ACS.

by cryo-TEM and by drop-casting on a carbon support grid shows that in toluene P3HT nanofibrils are formed in solution upon ageing prior to drop casting whereas in oDCB they tend to form rather during drop casting upon drying of the films because oDCB is a good solvent as compared to toluene. Interestingly, TEM reveals no significant difference in nanofibril dimensions in the cryo- and dried samples for both toluene and oDCB, indicating that the same type of nanofibrils of P3HT i.e. with the same crystalline form (form I) are generated in both solvents.

Remarkably, the authors managed to realize a 3D reconstruction of an individual P3HT nanofibril by electron tomography at high resolution. Figure 17(d) reveals the typical fringed patterns of P3HT nanofibrils with a characteristic periodicity of 1.7 nm corresponding to the packing of chains illustrated in Fig. 17(c). This packing is similar to that observed by HRTEM in

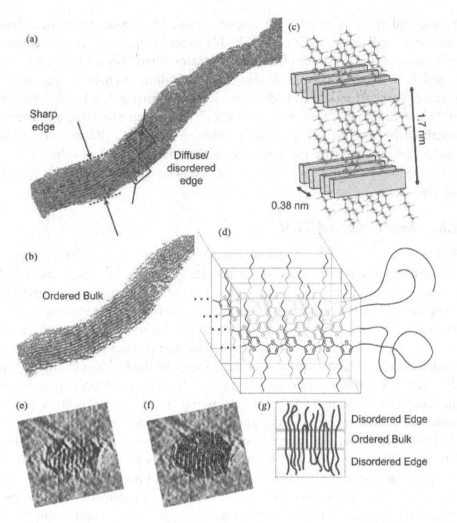

Figure 17: (a) Isosurface obtained from tomographic reconstruction of a P3HT nanofibril. (b) Isosurface from the middle of the nanofibril showing increased order. (c) Model of the lamellar order in the middle of the nanowire. (d) Model showing crystalline order in the bulk and disorder on the side of the nanofibril. (e) Slice from the tomographic reconstruction showing a cross-section of the nanofibril. (f) The same slice overlaid with the corresponding segment of the segmented structure. (g) Cartoon illustrating the ordered bulk and the disordered edges. (Reproduced with permission from Ref. 101 © 2014, ACS.

spin-coated films of P3ATs and shown in Figure 5. The reconstructed 3D image of a portion of the P3HT nanofibril indicates that the chains are highly ordered in the inner part of the nanofibril and tend to become disordered as they slowly branch out of the crystal (see Figs. 17(f) and 17(g)). In other words, the image obtained by cryo-tomography is consistent with both the model of fringed micelle and the model of chain folded crystal. In polyolefins, folded-chain lamellar crystals consist of a core of highly ordered chains which is decorated by an amorphous overlayer of disordered chains when they fold back into the crystal.[45] In the fringed micelle model, back folding of chains is minimal or absent. Recent structural considerations suggested that only a limited fraction of P3HT chains can fold back into the crystal given the large persistence length of this semi-flexible polymer (approx. 3 nm).[103,104] Importantly, these results demonstrate that the structure of nanofibrils grown in solution does not imply the formation of a clathrate i.e. a co-crystal between P3HT and the solvent.

4.3.   *Energy filtered-TEM*

Beside conventional BF TEM imaging, energy filtered TEM has become a tool of excellence for material's characterization over the last few decades.[29,30] When an electron beam goes through a thin film, the majority of electrons will pass unhindered leading to a so-called "zero-loss" signal whereas a given fraction of electrons does undergo elastic or inelastic scattering: these electrons can be used to characterize the material. The loss of momentum and energy is related to the elements present in the sample. In the low loss energy region ($E < 100$ eV), the inelastic interaction is related to plasmon generation in the sample as well as other mechanisms e.g. interband transitions and in some cases excitons.[29,30] From a practical point of view, the selection of the electrons with a certain energy can be done by using either an in-column filter or a post column spectrometer. Zero-loss imaging is of interest to image samples with low contrast such as biological probes. EFTEM has been used efficiently by various groups to understand in particular the structure in blends of organic semiconductors used in organic photovoltaic cells.[105,106] Figure 18 illustrates the case of 50/50 (w/w) blend film of P3HT and PCBM. The zero-loss image in Fig. 18(a) is formed from the elastically scattered electrons. The image is similar to a classical BF where the contrast originates essentially from differences in material density (1.1 g/cm^3 for P3HT and 1.5 g/cm^3 for PCBM). When the energy window is set to 19 eV, enhanced EELS intensity is observed from P3HT. The same area imaged at 30 eV shows essentially a complementary image where the PCBM-rich areas are bright. From these images, it becomes clear that P3HT forms a strongly interconnected network of nanofibrils surrounded by a PCBM-rich phase. The interconnected nature of this network is essential to channel out the holes generated after exciton dissociation at the PCBM/P3HT interface in the blend films.

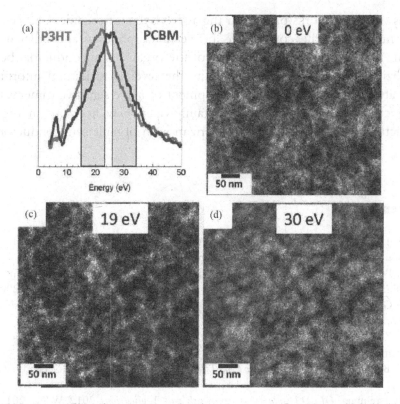

Figure 18: Energy filtered transmission electron microscopy of a 100 nm-thick 50:50 (w/w) blend of P3HT and PCBM: (a) ELS spectrum of pure P3HT and pure PCBM thin films. The energy windows in blue are for P3HT at 19 ± 4 eV and for PCBM at 30 ± 4 eV. (b) Zero-loss image formed only with elastically scattered electrons. (c) 19 eV loss image, where P3HT domains are bright. (d) 30 eV loss image, where PCBM domains are bright. (Reproduced with permission from Ref. 105 © American Chemical Society.)

5. Conclusion

This review has given a certain number of examples to illustrate the usefullness of TEM in the structural characterization of some representative molecular and polymeric semiconductors. The versatility of TEM has allowed significant progress in the understanding of the processing-structure-property *nexus*. "Old-fashioned" TEM methods such as electron diffraction in conjunction with controlled crystallization proved effective to address the crystalline packing of complex π-conjugated polymers such as P3HT or some low bandgap polymers. New observation modes such as STEM-HAADF or EFTEM have emerged over the last decade, opening new possibilities for material's characterization. TEM tomography made it possible to generate detailed morphological models of 3D networks present in the active layers of organic solar cells and better understand the correlations between morphology and efficiency. At present, the intrinsic beam-sensitivity of the organic semiconductors bearing solubilizing alkyl side chains still limits the level of structural information that is attainable by TEM. The development of more sensitive cameras based on direct-electron detection will certainly open new horizons in the field of structure and morphology characterization in organic semiconductor thin films.

References

1. A. J. Heeger, Nobel lecture: Semiconducting and metallic polymers: The fourth generation of polymeric materials. *Rev. Mod. Phys.*, **73**, 681–700 (2001).
2. J. Peet, M. L. Senatore, A. J. Heeger and G. C. Bazan, The role of processing in the fabrication and optimization of plastic solar cells. *Adv. Mat.*, **21**, 1521–1527 (2009).
3. S. Günes, H. Neugebauer, N. S. Sariciftci, Conjugated polymer-based organic solar cells. *Chem. Rev.*, **107**, 1324–1338 (2007).
4. S. Zaumseil and H. Sirringhaus, Electron and ambipolar transport in organic field-effect transistors. *Chem. Rev.*, **107**, 1296–1323 (2007).
5. Z. Bao and J. Locklin in *Organic Field Effect Transistors*, CRC Press, Boca Raton, (2007). p. 1.
6. T. Tsujimura, *OLED Displays Fundamentals and Applications*, 2012, Wiley, (2012) p. 1.
7. S. Holliday, J. E. Donaghey and I. McCulloch, Advances in charge carrier mobilities of semiconducting polymers used in organic transistors. *Chem. Mater.*, **26**, 647–663 (2014).
8. A. Facchetti, Semiconductors for organic transistors. *Mater. Today*, **10**, 28–37 (2007).
9. A. Salleo, Charge transport in polymeric transistors. *Mater. Today*, **10**, 38–45 (2007).
10. J. Peet, J. Y. Kim, N. E. Coates, W. L. Ma, D. Moses, A. J. Heeger and G. C. Bazan, Efficiency enhancement in low-bandgap polymer solar cells by processing with alkane dithiols. *Nat. Mater.*, **6**, 497–500 (2007).
11. M. Brinkmann, G. Gadret, M. Muccini, C. Taliani, N. Masciocchi et A. Sironi, Correlation between molecular packing and optical properties in different crystalline polymorphs and amorphous thin films of *mer*-Tris(8-hydroxyquinoline)aluminum(III). *J. Am. Chem. Soc.*, **122**, 5147–5157 (2000).
12. Y. Diao, B. C.-K. Tee, G. Giri, J. Xu, D. H. Kim, H. A. Becerril, R. M. Stoltenberg, T. H. Lee, G. Xue, S. C. B. Mannsfeld and Z. Bao, Solution coating of large-area organic semiconductor thin films with aligned single-crystalline domains. *Nat. Mat.* 12, 665–671 (2013).
13. R. J. Klein, M. D. McGehee and M. F. Toney, Highly oriented crystals at the buried interface in polythiophene thin-film transistors. *Nat. Mat.*, **5**, 222–228 (2006).
14. J. A. Lim, F. Liu, S. Ferdous, M. Muthukumar and A. L. Briseno, Polymer semiconductor crystals. *Mater. Today*, **13**, 14–24 (2010).
15. G. Yu, J. C. Hummelen, F. Wudl and A. J. Heeger, Polymer photovoltaic cells — enhanced efficiencies via a network of internal donor-acceptor heterojunctions: Enhanced

efficiencies via a network of internal donor-acceptor heterojunctions. *Science*, **270**, 1789–1791 (1995).

16. P. W. M. Blom, V. D. Mihailetchi, L. Jan Anton Koster and D. E. Markov, Device physics of polymer: Fullerene bulk heterojunction solar cells. *Adv. Mater.*, **19**, 1551–1566 (2007).

17. G. Renaud, R. Lazzari and F. Leroy Probing surface and interface morphology with grazing incidence small angle X-ray scattering. *Surf. Sci. Rep.*, **64**, 255–380 (2009).

18. P. Samori and J. P. Rabe, Scanning probe microscopy explorations on conjugated (macro)molecular architectures for molecular electronics. *J. Phys. Cond. Mat.*, **14**, 9955–9973 (2002).

19. H. Hoppe, T. Glatzel, M. Niggemann, A. Hinsch, M. C. Lux-Steiner and N. S. Sariciftci, Kelvin probe force microscopy study on conjugated polymer/fullerene bulk heterojunction organic solar cells. *Nano Letters*, **5**, 269–274 (2005).

20. J. Rivnay, S. C. B. Mannsfeld, C. E. Miller, A. Salleo and M. F. Toney, Quantitative determination of organic semiconductor microstructure from the molecular to device scale. *Chem. Rev.*, **112**, 5488–5519 (2012).

21. E. Barrena, D. G. De Oteyza, S. Sellner, H. Dosch, J. O. Osso and B. Struth, *in situ* study of the growth of nanodots in organic heteroepitaxy. *Phys. Rev. Lett.*, **97**, 076102 (2006).

22. W. Watts, T. Schuettfort and C. R. McNeill, Mapping of domain orientation and molecular order in polycrystalline semiconducting polymer films with soft X-ray microscopy. *Adv. Funct. Mater.*, **21**, 1122–1131 (2011).

23. B. Watts, C. R. McNeill, J. Raabe, Imaging nanostructures in organic semiconductor films with scanning transmission X-ray spectro-microscopy. *Synth. Met.* 161, 2516–2520 (2012).

24. P. B. Hirsch, R. B.B Nicholson, A. Howie, D. W. Pashley and M. J. Whelan, *Electron Microscopy of Thin Crystals*, Butterworth & Co LTD, (1965), p. 1–23.

25. D. B. Williams and C. B. Carter, *Transmission Electron Microscopy, a Textbook for Materials Science*, Springer Science+Business Media, (1996), p. 1.

26. G. L. Mischler, *Electron Microscopy of Polymers*, Springer Verlag, Berlin Heidelberg, 2008. p. 1.

27. M. R. Libera and R. F. Egerton, Advances in the transmission electron microscopy of polymers. *Polym. Rev.*, **50**, 2010, 321–339 (2010).

28. T. Kobayashi, *Epitaxial Growth of Organic Thin Films and Characterization of Their Defect Structures by High-resolution Electron Microscopy*, Edited by D. H. C., Freyhardt and G. Müller. *Crystals, Growth, Properties and Applications*, Springer Verlag, (1991), p. 1.

29. R. Brydson, *Electron Energy Loss Spectroscopy*. New York, Taylor & Francis, (2001), p. 1.

30. R. F. Egerton, Electron energy-loss spectroscopy in the TEM. *Rep. Prog. Phys.*, **72**, 1 (2008).

31. X. Yang and J. Loos, Toward high-performance polymer solar cells: The importance of morphology control. *Macromolecules*, **40**, 1353–1362 (2007).

32. M. M. Freundlich, Origin of the electron microscope. *Science*, **142**, 185–188 (1963).

33. R. H. Geiss, Electron diffraction of polymers. *Encyclopedia of Polymer Science and Technology*, 1–23 (2014).

34. S. J. Pennycook, P. D. Nellist, *Scanning Transmission Electron Microscopy: Imaging and Analysis*. Springer, (2011) p. 1.

35. J. Loos, E. Sourty, K. Lu, G. de With and S. v. Bavel, Imaging polymer systems with high-angle annular dark field scanning transmission electron microscopy (HAADF–STEM). *Macromolecules*, **42**, 2581–2586 (2009).

36. R. F. Egerton, P. Li and M. Malac, Radiation damage in the TEM and SEM. *Micron*, **35**, 399–409 (2004).

37. R. F. Egerton, Control of radiation damage in the TEM. *Ultramicroscopy*, **127**, 100–108 (2013).

38. D. Dorset, *Structural Electron Crystallography*. 1995, Plenum Press, New York, p. 1.

39. M. Brinkmann, Structure and morphology control in thin films of regioregular poly(3-hexylthiophene). *J. Polym. Sci. B, Polymer Phys.*, **49**, 1218–1233 (2011).

40. M. Brinkmann and J.-C. Wittmann, Orientation of regioregular poly(3-hexylthiophene) by directional solidification: A simple method to reveal the semicrystalline structure of a conjugated polymer. *Adv. Mat.*, **18**, 860–863 (2006).

41. M. Brinkmann and P. Rannou, Effect of the molecular weight on the structure and morphology of oriented thin films of regioregular poly(3-hexylthiophene) grown by directional epitaxial crystallization. *Adv. Funct. Mat.*, **17**, 101–108 (2007).

42. M. Brinkmann, L'épitaxie des polymères conjugués semi-conducteurs. *L'actu. Chim.*, **326**, 31–34 (2009).

43. M. Brinkmann, L. Hartmann, N. Kayunkid and D. Djurado, Understanding the structure and crystallization of regioregular poly (3-hexylthiophene) from the perspective of epitaxy. *Advances in Polym. Sci.*, **265**, 83–106 (2014).

44. J. Petermann and R. M. Gohil, A new method for the preparation of high modulus thermoplastic films. *J. of Mater. Sci.*, **14**, 2260–2263 (1979).

45. B. Lotz and J.C. Wittmann, Chapter 3, Structure of polymer single crystals, Edited by R. W. Cahn, P. Haasen, E. J. Kramer, *Materials Science and Technology: A Comprehensive Treatment*. Vol. 12. Weinheim, V.C.H. (1993), p. 79–151.

46. M. Brinkmann, Directional epitaxial crystallization and tentative crystal structure of poly(9,9′-di-*n*-octyl-2,7-fluorene). *Macromolecules*, **40**, 7532–7541 (2007).

47. M. Brinkmann, N. Charoenthai, R. Traiphol, P. Piyakulawat, J. Wlosnewski and U. Asawapirom, Structure and morphology in highly oriented films of poly(9,9-bis(*n*-octyl)fluorene-2,7-diyl) and poly(9,9-bis(2-ethylhexyl)fluorene-2,7-diyl) grown on friction transferred poly(tetrafluoroethylene). *Macromolecules*, **42**, 8298–8306 (2009).

48. G. Lieser, M. Oda, T. Miteva, A. Meisel, H.-G. Nothofer and U. Scherf, Ordering, graphoepitaxial orientation, and conformation of a polyfluorene derivative of the "hairy-rod" type on an oriented substrate of polyimide. *Macromolecules*, **33**, 4490–4495 (2000).

49. N. Uyeda, T. Kobayashi, K. Ishizuka, Y. Fujiyoshi, Crystal structure of Ag ·TCNQ. *Nature*, **285**, 95–97 (1980).

50. J. R. Fryer, High-resolution imaging of organic crystals. *J. Elect. Microsc. Tech.*, **11**, 310–325 (1989).

51. J. C. H. Spence, *High-Resolution Electron Microscopy*. Oxford University Press, (2003). p. 1.

52. L. F Drummy and D. C. Martin, Thickness-driven orthorhombic to triclinic phase transformation in pentacene thin films. *Adv. Mater.*, **17**, 903–907 (2005).

53. M. Brinkmann, S. Graff, J.-C. Wittmann, C. Chaumont, F. Nuesch, A. Anver, M. Schaer and L. Zuppiroli, Orienting tetracene and pentacene thin films onto friction-transferred poly(tetrafluoroethylene) substrate. *J. Phys. Chem. B*, **107**, 10531–10539 (2003).

54. G. Horowitz and M. E. Hajlaoui, Mobility in polycrystalline oligothiophene field-effect transistors dependent on grain size. *Adv. Mater.*, **12**, 1046–1050 (2000).

55. J. Chen, C. K. Tee, M. Shtein, J. Anthony and D. C. Martin, Grain-boundary-limited charge transport in solution-processed 6,13 bis(tri-isopropylsilylethynyl) pentacene thin film transistors. *J. Appl. Phys.*, **103**, 114513 (2008).

56. D. J. Gundlach, Y. Y. Lin, T. N. Jackson and S. F. Nelson, Pentacene organic thin-film transistors-molecular ordering and mobility. *IEEE Elect. Device Lett.*, **18**, 87–89 (1997).

57. C. C. Mattheus, G. A. De Wijs, R. A. De Groot and T. T. M. Palstra, Modeling the polymorphism of pentacene. *J. Am. Chem. Soc.*, **125**, 6323–6330 (2003).

58. M. Brinkmann and P. Rannou, Molecular weight dependence of chain packing and semicrystalline structure in oriented films of regioregular poly(3-hexylthiophene) revealed by high-resolution transmission electron microscopy. *Macromolecules*, **42**, 1125–1130 (2009).

59. S. T. Salammal, E. Mikayelyan, S. Grigorian, U. Pietsch, N. Koenen, U. Scherf, N. Kayunkid, M. Brinkmann, Impact of thermal annealing on the semicrystalline nanomorphology of spin-coated thin films of regioregular poly(3-alkylthiophene)s as observed by high-resolution transmission electron microscopy and grazing incidence X-ray diffraction. *Macromolecules*, **45**, 5575–5585 (2012).

60. A. Hamidi Sakr, L. Biniek, S. Fall and M. Brinkmann, Precise control of lamellar thickness in highly oriented regioregular poly(3-hexylthiophene) thin films prepared by high-temperature rubbing: Correlations with optical properties and charge transport. *Adv. Funct. Mat.*, **26**, 408–420 (2016).

61. Y. Sun, G. C. Welch, W. L. Leong, C. J. Takacs, G. C. Bazan and A. J. Heeger Solution-processed small-molecule solar cells with 6.7% efficiency. *Nature Mat.*, **11**, 44–48 (2012).

62. J.-F. Moulin, M. Brinkmann, A. Thierry and J.-C. Wittmann, Oriented crystalline films of tris(8-hydroxyquinoline) Aluminum(III): Growth of the alpha polymorph onto an ultra-oriented poly(tetrafluoroethylene) Substrate. *Adv. Mat.*, **14**, 436–439 (2002).

63. M. Brinkmann, B. Fite, S. Pratontep et C. Chaumont, Structure and spectroscopic properties of the crystalline structures containing meridional and facial isomers of tris(8-hydroxyquinoline) Gallium(III). *Chem. Mat.*, **16**, 4627–4633 (2004).

64. C. Vergnat, J.-F. Legrand and M. Brinkmann, Orienting semiconducting nanocrystals on nanostructured polycarbonate substrates: Impact of substrate temperature on polymorphism and in-plane orientation. *Macromolecules*, **44**, 3817–3827 (2011).

65. M. Brinkmann, J.-C. Wittmann, M. Barthel, M. Hanack et C. Chaumont, Highly ordered titanyl phthalocyanine films grown by directional crystallization on oriented poly(tetrafluoroethylene) substrate. *Chem. Mat.*, **14**, 904–914 (2002).

66. C. Vergnat, S. Uttiya, S. Pratontep, J.-F. Legrand and M. Brinkmann, Oriented growth of zinc(II) phthalocyanines on polycarbonate alignment substrates: Effect of substrate temperature on the in-plane orientation. *Synth. Met.*, **161**, 251–258 (2011).

67. C. Vergnat, V. Landais, J. Combet, A. Vorobiev, O. Konovalov, J.-F. Legrand and M. Brinkmann, Comparing the growth of a molecular semiconductor on amorphous and semi-crystalline polycarbonate substrates. *Organ. Elect.*, **13**, 1594–1601 (2012).

68. T. Djuric, G. Hernandez-Sosa, G. Schwabegger, M. Koini, G. Hesser, M. Arndt, M. Brinkmann, H. Sitter, C. Simbrunner and R. Resel, Alternately deposited heterostructures

of α-sexithiophene–*para*-hexaphenyl on muscovite mica(001) surfaces: crystallographic structure and morphology. *J. Mat. Chem.*, **22**, 15316–15325 (2012).

69. T. Djuric, T. Ules, S. Gusenleitner, N. Kayunkid, H. Plank, G. Hlawacek, C. Teichert, M. Brinkmann, M. Ramsey and R. Resel, Substrate selected polymorphism of epitaxially aligned tetraphenyl-porphyrin thin films. *Phys. Chem. Chem. Phys.*, **14**, 262–272 (2012).

70. C. Simbrunner, G. Hernandez-Sosa, M. Oehzelt, T. Djuric, I. Salzmann, M. Brinkmann, G. Schwabegger, I. Watzinger, H. Sitter, and R. Resel, Epitaxial growth of sexithiophene on mica surfaces. *Phys. Rev. B*, **83**, 115443 (2011).

71. H. Plank, R. Resel, S. Purger, J. Keckes, A. Thierry, B. Lotz, A. Andreev, N. S. Sariciftci, and H. Sitter, Heteroepitaxial growth of self-assembled highly ordered *para*-sexiphenyl films: A crystallographic study. *Phys. Rev. B*, **64**, 235423 (2001).

72. L. Biniek, P.-O. Schwartz, E. Zabrova, B. Heinrich, N. Leclerc, S. Méry, M. Brinkmann, Zipper-like molecular packing of donor–acceptor conjugated co-oligomers based on perylenediimide. *J. Mater. Chem. C*, **3**, 3342–3349 (2015).

73. Y. J. Park, S. J. Kang, B. Lotz, M. Brinkmann, A. Thierry, K. Kim, J. C. Park, Ordered ferroelectric PVDF−TrFE thin films by high throughput epitaxy for nonvolatile polymer memory. *Macromolecules*, **41**, 8648–8654 (2008).

74. M. Brinkmann, C. Contal, N. Kayunkid, T. Djuriç, R. Resel, Nanotexturation of a conjugated semi-conducting polymer by epitaxial growth on an aromatic substrate. *Macromolecules*, **43**, 7604–7610 (2010).

75. B. Moss and D. L. Dorset, Refinement of linear polymer crystal structures determined from electron diffraction data. *J. Polym. Sci. B: Polym. Phys.*, **20**, 1789–1804 (1982).

76. B. Lotz, Frustration and frustrated crystal structures of polymers and biopolymers. *Macromolecules*, **45**, 2175–2189 (2012).

77. L. Cartier, N. Spassky and B. Lotz, Crystal structure of poly(*tert*-butylethylene sulfide): A reappraisal in the light of frustration. *Macromolecules*, **31**, 3040–3048 (1998).

78. L. Cartier, T. Okihara and B. Lotz, The α "Superstructure" of syndiotactic polystyrene: A frustrated structure. *Macromolecules*, **31**, 3303–3310 (1998).

79. K. Rahimi, I. Botiz, N. Stingelin, N. Kayunkid, M. Sommer, F. Koch, H. Nguyen, O. Coulembier, P. Dubois, M. Brinkmann and G. Reiter, Controllable processes for generating large single crystals of poly(3-hexylthiophene). *Angew. Chem. Int. Ed.*, **51**, 11131–11135 (2012).

80. A. Buono, N. H. Son, G. Raos, L. Gila, A. Cominetti, M. Castellani and S. V. Meille, Form II poly(3-butylthiophene): Crystal structure and preferred orientation in spherulitic thin films. *Macromolecules*, **43**, 6772–6781 (2010).

81. N. Kayunkid, S. Uttiya and M. Brinkmann, Structural model of regioregular poly(3-hexylthiophene) obtained by electron diffraction analysis. *Macromolecules*, **43**, 4961–4967 (2010).

82. F. S. U. Fischer, N. Kayunkid, D. Trefz, S. Ludwigs and M. Brinkmann, Structural models of poly(cyclopentadithiophene-*alt*-benzothiadiazole) with branched side chains: impact of a single fluorine atom on the crystal structure and polymorphism of a conjugated polymer. *Macromolecules*, **48**, 3974–3982 (2015).

83. L. Biniek, N. Leclerc, T. Heiser, M. Brinkmann, Large scale alignment and charge transport anisotropy of pBTTT films oriented by high temperature rubbing. *Macromolecules*, **46**, 4014–4023 (2013).

84. K. Tremel, F. S. U. Fischer, N. Kayunkid, R. DiPietro, A. Kiriy, D. Neher, S. Ludwigs, M. Brinkmann, Charge transport anisotropy in highly oriented thin films of the acceptor polymer P(NDI2OD-T2). *Advanced Energy Materials*, **4**, 1301659 (2014).

85. L. Biniek, S. Poujet, D. Djurado, E. Gonthier, K. Tremel, N. Kayunkid, E. Zaborova, N. Crespo-Monteiro, O. Boyron, N. Leclerc, S. Ludwigs and M. Brinkmann. High-temperature rubbing: A versatile method to align π-conjugated polymers without alignment substrate. *Macromolecules*, **47**, 3871–3879 (2014).

86. M. Brinkmann, E. Gonthier, S. Bogen, K. Tremmel, S. Ludwigs, M. Hufnagel, M. Sommer, Segregated *versus* mixed interchain stacking in highly oriented films of naphthalene diimide bithiophene copolymers. *ACS Nano*, **6**, 10319–10326 (2012).

87. C. Liu, K. Wang, X. Gong and A. J. Heeger, Low bandgap semiconducting polymers for polymeric photovoltaics. *Chem. Soc. Rev.*, **2016**, in press.

88. X. Zhan, A. Facchetti, S. Barlow, T. J. Marks, M. A. Ratner, M. R. Wasielewski and S. R. Marder, Rylene and related diimides for organic electronics. *Adv. Mat.*, **23**, 268–284 (2011).

89. M. Sommer, Conjugated polymers based on naphthalene diimide for organic electronics. *J. Mater. Chem. C*, **2**, 3088–3098 (2014).

90. M. Weyland. and P. A. Midgley, Electron tomography. *Materials Today*, **7**, 32–40 (2004).

91. M. Radermacher, Three-dimensional reconstruction of single particles from random and nonrandom tilt series. *J. Electron Microsc. Tech.*, **9**, 359–394 (1988).

92. G. Möbus, and B. J.Inkson, Nanoscale tomography in materials science. *Materials Today*, **10**, 18–25 (2007).

93. A. J. Koster, U. Ziese, A. J. Verkleij, A. H. Janssen and K. P. de Jong, Three-dimensional transmission electron microscopy: A novel imaging and characterization technique with nanometer scale resolution for materials science. *J. Phys. Chem. B*, **104**, 9368–9370 (2000).

94. Y. Huang, E. J. Kramer, A. J. Heeger and G. C. Bazan, Bulk heterojunction solar cells: Morphology and performance relationships. *Chem. Rev.*, **114**, 7006–7043 (2014).

95. S. S. van Bavel, E. Sourty, G. de With and J. Loos, Three-dimensional nanoscale organization of bulk heterojunction polymer solar cells. *Nano. Lett.*, **9**, 507–513 (2009).

96. S. S. Van Bavel, E. Sourty, G. de With, S. Veenstra and J. Loos, Three-dimensional nanoscale organization of polymer solar cells. *J. Mater. Chem.*, **19**, 5388–5393 (2009).

97. S. S. Van Bavel, M. Bärenklau, G. De With, H. Hoppe and J. Loos, P3HT/PCBM bulk heterojunction solar cells: Impact of blend composition and 3D morphology on device performance. *Adv. Funct. Mater.*, **20**, 1458–1463 (2010).

98. S. Samitsu, T. Shimomura, S. Heike, T. Hashizume and K. Ito, Effective production of poly(3-alkylthiophene) nanofibers by means of whisker method using anisole solvent: Structural, optical, and electrical properties. *Macromolecules*, **41**, 8000–8010 (2008).

99. J. Y. Oh, M. Shin, T. I. Lee, W S. Jang, Y. Min, J.-M. Myoung, H. K. Baik, and U. Jeong, Self-seeded growth of poly(3-hexylthiophene) (P3HT) nanofibrils by a cycle of cooling and heating in solutions. *Macromolecules*, **45**, 7504–7513 (2012).

100. D. Danino, Cryo-TEM of soft molecular assemblies. *Current. Opin. Coll. & Interf. Sci.*, **17**, 316–329 (2012).

101. M. J. M. Wirix, P. H. H. Bomans, M. M. R. M. Hendrix, H. Friedrich, N. A. J. M. Sommerdijk and G. de With, Three-dimensional structure of P3HT assemblies in organic solvents revealed by cryo-TEM. *Nanoletters*, **14**, 2033–2038 (2014).
102. M. J. M. Wirix, P. H. H. Bomans, H. Friedrich, N. A. J. M. Sommerdijk and G. de, With visualizing order in dispersions and solid state morphology with Cryo-TEM and electron tomography: P3HT:PCBM organic solar cells. *J. Mat. Chem. A*, **3**, 5031–5040 (2015).
103. C. R. Snyder, R. C. Nieuwendaal, D. DeLongchamp, C. K. Luscombe, P. Sista and S. D. Boyd, Quantifying crystallinity in high molar mass poly(3-hexylthiophene). *Macromolecules*, **47**, 3942–3950 (2014).
104. B. McCullough, V. Ho, M. Hoarfrost, C. Stanley, C. Do, W. T. Heller and R. A. Segalman, Polymer chain shape of poly(3-alkylthiophenes) in solution using small-angle neutron scattering. *Macromolecules*, **46**, 1899–1907 (2013).
105. L. F. Drummy, R. J. Davis, D. L. Moore, M. Durstock, R. Vaia and J. W. P. Hsu, Molecular-scale and nanoscale morphology of P3HT:PCBM bulk heterojunctions: energy-filtered TEM and low-dose HREM. *Chem. Mater.*, **23**, 907–912 (2011).
106. S. Albrecht, S. Janietz, W. Schindler, J. Frisch, J. Kurpiers, J. Kniepert, S. Inal, P. Pingel, K. Fostiropoulos, N. Koch, D. Neher, Fluorinated copolymer PCPDTBT with enhanced open-circuit voltage and reduced recombination for highly efficient polymer solar cells. *J. Am. Chem. Soc.*, **134**, 14932–14944 (2012).

Chapter 3

Structuring Conjugated Polymers and Polyelectrolytes Through Self-Assembly

Hugh D. Burrows*, Beverly Stewart,
M. Luisa Ramos and Licinia L. G. Justino

Centro de Química, Department of Chemistry,
University of Coimbra, 3004-535. Coimbra, Portugal
* burrows@ci.uc.pt

This chapter considers the use of self-assembly to nanostructure conjugated polymers and related systems. Self-assembly involves interactions such as hydrogen polymers, electrostatic interactions, van der Waals, steric repulsions, and the hydrophobic/hydrophilic balance. Specific experimental articles are discussed, including photofunctional nanoparticles from zinc 5,10,15,20-tetrakis(perylenediimide)-porphyrin building blocks, a zwitterionic thiophene-based conjugated polymer interacting with surfactants, and aqueous solutions of poly[9,9-bis(4-sulfonylbutoxyphenyl)fluorene-2,7-diyl-2',2'-bithiophene] (PBS-PF2) in the presence of Pentaethylene Glycol Monododecyl Ether ($C_{12}E_5$). Particular emphasis is given to information obtained from nuclear magnetic resonance (NMR) spectroscopy and various modelling and computational studies, such as ultra-high aspect ratio nanowires, H-aggregates, interaction of 8-hydroxyquinoline-5-sulfonate with metal ions and surfactants, β-phase formation in polyfluorenes, effects involving π-stacking and hydrogen bonding. It is shown that the combination of information from structural studies using techniques such as NMR spectroscopy with theoretical methodologies, ranging from *ab initio* and density functional theory (DFT) calculations to molecular dynamics simulations (MDSs), is starting to suggest guidelines for the specific design of advanced materials for various electronic and optoelectronic applications.

1. Introduction

The increasing application of conjugated polymers as organic semiconductors in areas such as polymer light emitting diodes (PLEDs), solar cells, chemical and biological sensors is due in part to the possibility of processing them from solution by techniques, such as inkjet and screen printing. It is in fact possible to produce large area devices cheaply using roll-to-roll methodologies. In addition, devices can be produced on flexible supports. This has led to considerable interest into the study of their solution behaviour. By extension it has stimulated the development of conjugated polyelectrolytes (CPEs), conjugated polymers containing ionic groups that are soluble in water or other polar solvents. However, these often tend to aggregate in solution, which can affect properties, such as luminescence yields when used in devices.[1] However, it is possible to overcome this aggregation using cosolvents, surfactants or water soluble polymers, and there is considerable interest in understanding how to control such processes.

Hybrid systems containing conjugated polymers and small molecules are of interest for applications, such as heavy metal complex doped phosphorescent PLEDs[2] and bulk heterojunction photovoltaic devices.[3] However, such hybrid systems have a tendency to phase separate.[4] It has been shown that this can be avoided by using electrostatic interactions between oppositely charged CPEs and small molecules, either for light emitting complexes for LEDs and light harvesting[5,6] or CPE-fullerene systems for application in solar cells.[7] Such electrostatic interactions form part of a wider range of noncovalent interactions which can be used to self-assemble molecules on a wide range of length scales to produce supramolecular structures.[8–11] This can be used for nanostructuring to produce materials involving conjugated polymers for a wide range of applications.[12,13] In this chapter we will provide a general discussion of self-assembly, and will then present examples of such systems, and of theoretical and other ways of studying and modelling them.

2. Types on Interactions Involved in Self-Assembly

The investigation of specific interactions which occur when mixtures of species undergo spontaneous self-assembly is an issue worthy of considerable attention. It is these intercommunications between species which are responsible for the shape and type of molecular structures formed and, as a direct consequence, the applications of such systems.

Self-assembly occurs through a range of intermolecular and interatomic interactions, examples of such are hydrogen bonding, electrostatic interactions, van der Waals interactions, steric repulsions, as well as those more related to bulk systems such as hydrophobic/hydrophilic behaviours. There are many investigations into the influence these interactions have on phase formation and spontaneous molecular self-assembly. Here examples of the importance of interactions are discussed in terms of their significance on self-assembly and how a detailed study of these systems is crucial to greater understanding of the phenomena.

2.1. *Charge transport in photofunctional nanoparticles self-assembled from zinc 5,10,15,20-tetrakis (perylenediimide)-porphyrin building blocks*

Van der Waals forces, the residual attractive or repulsive forces between molecules or atomic groups which do not arise from covalent or ionic bonds, are an important factor in self-assembly. van der Boom *et al.*[14] show that van der Waals forces are responsible for driving self-assembly in ordered nanoparticles, both in solution and also in the solid state and this work will be discussed here.

To successfully design and develop molecular devices for application in photochemical energy conversion requires extensive research into the different physical behaviours that occur on nm scale distances. Artificial photosynthetic systems[15,16] as well as single molecules for use as switches and wires have been the subject of intensive study.[17–28] In recent years several strategies have been developed to control charge transfer in organic nanostructures that utilize the speed and efficiency of ultrafast photoinduced electron-transfer.[29,30] An important step towards photofunctional devices is to create increasingly large arrays of interactive molecules.

van der Boom *et al.* use the fact that covalent synthesis of such large molecules is prohibitive due to inefficiency and cost as the motivation for their work and discuss how, as a result of the exhaustive covalent approach, self-assembly is the preferred method by which to achieve ordered systems from smaller functional components. Self-assembly comprises a variety of weak interactions such as hydrogen bonding,[31–34] π-π and van der Waals interactions,[35,36] as well as covalent bonds[37–39] occurring between species resulting in the spontaneous formation of ordered aggregates.

The most important model for efficient photochemical energy conversion is photosynthesis where the cross section for the absorption of light is provided by many chlorophylls contained in antenna proteins.[40] Light absorption by antenna chlorophyll creates singlet excitons which migrate rapidly throughout the chlorophyll. Some of these excitons are eventually trapped by the primary

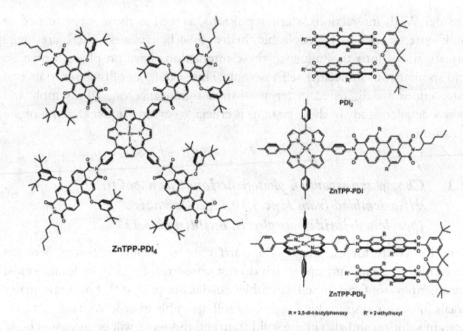

Figure 1:   zinc 5,10,15,20-tetrakis(perylenediimide)-porphyrin, ZnTPP-PDI$_4$ (Reproduced with permission from Ref. 14 Copyright 2002 American Chemical Society.)

donor chlorophyll within a nearby reaction centre protein leading to charge separation of a few picoseconds.

The study reports a zinc 5,10,15,20-tetrakis(perylenediimide)-porphyrin, ZnTPP-PDI$_4$, Fig. 1, which self assembles into ordered photofunctional organic nanoparticles using van der Waals interactions between the individual molecules and not only exhibits the characteristics of an antenna reaction center but, additionally, demonstrates more extensive charge transport within the nanoparticles as well.

The mass spectrum of ZnTpp-PDI$_4$ obtained using MALDI/TOF techniques shows peaks at multiples of the m/z = 4310 parent ion up to around 80 kD and is determined to relate to assemblies of around 20 molecules, thus providing strong evidence for the tendency of ZnTPP-PDI$_4$ to self assemble in the gas phase. Dynamic light-scattering measurements on a 10^{-4} M solution of ZnTPP-PDI$_4$ in toluene additionally show that this molecule also forms large soluble nanoparticles with a 150 ± 30 nm diameter, assuming a spherical particle model. The nanoparticle size was found to decrease with decreasing concentration, reaching 40 nm at 10^{-5} M, which was the limit of the light-scattering measurements used in the study.

Further evidence for self-assembly in solution to form (ZnTPP-PDI$_4$)$_n$ nanoparticles is also shown by proton NMR, where broadened lines in chloroform, dimethyl sulfoxide, and pyridine at temperatures up to 100°C and at concentrations down to 10^{-5} M are observed. In contrast, the NMR and mass spectra of ZnTPP, ZnTPP-PDI, and ZnTPP-PDI$_2$ showed that these molecules do not self-associate in solution.

The optical absorption spectrum of ZnTPP-PDI in toluene shows the ZnTPP B-band at 424 nm and the lowest energy Q-band at 592 nm, with an intense band at 550 nm due to the PDI molecule with its lower intensity vibronic band at 509 nm. Comparison of the PDI bands within ZnTPP-PDI to those of the (ZnTPP-PDI$_4$)$_n$ nanoparticles showed the ZnTPP B bands are very similar, while the (ZnTPP-PDI$_4$)$_n$ nanoparticles exhibit a new intense 515 nm PDI band. Due to the fact the transition dipoles of two identical chromophores are positioned in a parallel, stacked geometry, the molecular exciton model was applied to predict that coupling of the two transition dipoles would cause the lowest energy electronic transition of the dimer to split into two bands, with the higher energy band having all of the oscillator strength.[41] The application of the model was used to account for the presence of the intense 515 nm band in (ZnTPP-PDI$_4$)$_n$ and proposes that the PDI molecules in this nanoparticle adopt a parallel, stacked geometry, while the lack of a significant bandshift and/or splitting for the ZnTPP molecules suggests that they have a relative distance and/or geometry in which exciton coupling is weak. Further investigation theorizes that the PDI molecules most likely stack directly on top of one another presumably at a van der Waals distance of ∼3.5 Å while the Zn-TPP molecules occupy sites in every other layer with an interlayer Zn-Zn distance of around 7 Å, Fig. 2.

It was concluded from the study that self-assembly of ZnTPP-PDI$_4$ into (ZnTPP-PDI$_4$)$_n$ nanoparticles is driven primarily by strong van der Waals interactions in PDI molecules. The PDI chromophores provide structures that are analogous to the roles of chlorophyll within photosynthetic proteins by providing dual antenna and charge carrier functions. Additionally it was

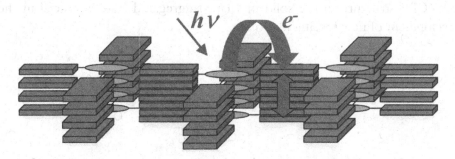

Figure 2: Schematic of the structure of self assembled (ZnTTP-PDI$_4$)$_{12}$ reproduced with permission from Ref. 14 Copyright 2002 American Chemical Society.

found that photoexcitation of these nanoparticles initiates an ultrafast charge separation that occurs with near unit efficiency. However, this results in an electron migration between a number of closely coupled electron acceptors which is a process more closely related to dye sensitised charge injection into a semiconductor than to a stepwise photosynthetic electron transfer.

2.2.   *Interactions of a zwitterionic thiophene-based conjugated polymer with surfactants*

Water soluble conjugated polythiophene based polyelectrolytes (PT) have been revealed as a class of molecules worthy of attention as active materials in optoelectronic devices and also biosensing.[42] Polythiophene derivatives possess distinctive chromic features which manifest as colour changes arising from changes in π-π^* transitions. These transition changes are related to conformational fluctuations in the backbone of the polymer from planar to nonplanar.[43–51]

In an investigation of the effects of various surfactants on a zwitterionic conjugated polyelectrolyte (ZCPE) poly[3-(N-(4-sulfonato-1-butyl)N,N-diethylammonium)hexyl-2,5-thiophene] (P3SBDEAHT), Fig. 3, it was shown that different surfactants have an influence on the formation of consequent P3SBDEAHT:surfactant aggregates.[52] The focus of the work carried out relates to changing the electronic environment to examine the effect this has on polymer conformation. ZCPEs show promise for applications as sensing platforms, for example a polar polythiophene carrying amino acid side chains (POWT) can interact with positively or negatively charged peptides, as well as with four-helix bundles formed by the two peptide chains.[53] Different polythiophene–peptide complexes are formed, with characteristic optical features, where the differences are due to changes in the effective conjugation length and/or due to interchain interactions and aggregation effects. Similar effects are also found when POWT interacts with ss-DNA or ds-DNA.[54,55] Hence, a detailed characterization of the ZCPE structure in the solid or (nano)aggregated state is crucial in the development of new sensing platforms.

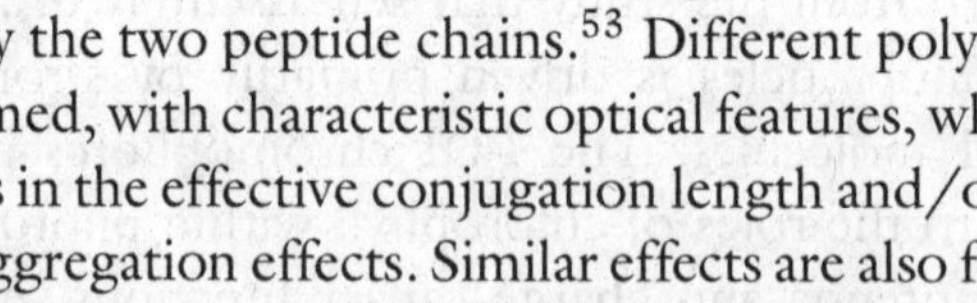

Figure 3:   Zwitterionic conjugated polythiophene P3SBDEAHT.

In this study by Costa *et al.*[52] the behavior of the zwitterionic P3SBDEAHT in water, and with various types of surfactants was examined. Analysis of optical and structural properties of P3SBDEAHT in water with a range of surfactants was carried out using a variety of techniques such as absorption, photoluminescence (PL), electrical conductivity, molecular dynamics simulations (MDSs) and small-angle X-ray scattering (SAXS). It revealed that complex formation is influenced by both non-specific and also electrostatic interactions. These interactions are responsible for the spatial for the spatial position of the surfactant in relation to P3SBDEAHT within the complex.

Zwitterionic P3SBDEAHT in aqueous solution is characterized by a non-resolved absorption band peaking at 445 nm and a broad featureless emission band peaking at 600 nm, in mixtures of water with a miscible organic solvent such as dioxane both emission intensity and position of the emission band maximum change with respect to the solute. Dioxane presence leads to an increase in PL intensity and a blue shift of the PL band. When dioxane content is around 30–50% PL intensity is found to be higher with a concomitant blue shift relative to pure water or high dioxane content. At higher dioxane content the intensity decreases and a reverse red-shift is observed. These shifts and PL changes are explained in terms of intermolecular interactions. It has been demonstrated previously that solvent effects on PL spectra are associated with the changes in aggregation state of the solute due to different balances between hydrophobic and hydrophilic interactions in mixtures.[56–60] In water interchain hydrophobic interactions (π-π interactions) are dominant with aggregates forming to avoid contact between the hydrophobic chains of the polymer and water. This proximity between the conjugated segments of different polymer chains allows interchain migration of photoexcitation and consequently emission from the lower-energy emitting sites occurs resulting in lower photoluminescence and red-shifted emission.

In 30–50% dioxane mixtures, the optimum balance between hydrophobic and hydrophilic interactions is attained; interchain interactions are excluded and the PL intensity increases. At higher dioxane concentrations new aggregates are formed. A decrease of the polarity of the solvent favours polymer backbone-solvent interaction and consequently P3SBDEAHT orients itself in order to avoid less favourable solvent-zwitterionic side chain interactions and a red-shift is again observed.

Similar effects were observed with the addition of surfactants to an aqueous solution of P3SBDEAHT. Effects of non-ionic (S^0), anionic (S^-), cationic (S^+) and zwitterionic ($S^{-/+}$) surfactant were studied and different behaviours were observed that were dependent on the nature of the surfactant added.

Addition of cationic surfactants, dodecyltrimethylammonium chloride (DTAC) and hexadecyltrimethylammonium bromide (CTAB) cause similar effects on the optical properties of P3SBDEAHT. Below the critical micelle concentration (*cmc*) of the cationic surfactant, the intensity, the maximum and the shape of the PL spectra does not significantly change, however, when the surfactant concentration approaches the *cmc*, an increase of PL intensity and a blue-shift (by ca. 6 nm) of the spectra is identified. The surfactant concentration at which this change is observed corresponds to the critical aggregation concentration (*cac*). This concentration is, most probably, due to conformational changes of single P3SBDEAHT chains in P3SBDEAHT/anionic surfactant aggregates.

Anionic sulfate surfactants (with chain lengths n = 8, 12 and 14, labelled as SOS, SDS and STS, respectively) were found to induce different changes in the optical spectra of P3SBDEAHT when compared with cationic surfactants. Upon addition of the anionic surfactant (*csurf* ≪ *cmc*) (where *csurf* is the surfactant concentration at which the maximum absorption (λ_{abs}) and emission (λ_{em}) wavelengths, and fluorescence quantum yields (ϕ_F) of P3SBDEAHT in water and in aqueous micellar solutions was obtained), the PL intensity is first enhanced and the PL band blue-shifted. The PL intensity continuously increases upon further addition of the surfactant. Starting at concentrations close to the *cmc* the PL intensity and emission maximum remain constant during further surfactant addition. Two switching points for the dependence of PL intensity and λ_{max} vs. surfactant concentration were identified. The first break occurring at concentrations well below the *cmc* and may be due to some disaggregation of interchain aggregates. The second turn corresponding to the critical aggregation constant (*cac*) was also observed in P3SBDEAHT/S$^+$ surfactant systems.

Comparing the optical responses of aqueous P3SBDEAHT solutions in the presence of single-tail surfactants with dodecyl chains (anionic SDS, cationic DTAC, and the non-ionic surfactant $C_{12}E_5$) the surfactochromic effect[51] was found to increase in the following order: $C_{12}E_5$ < DTAC < SDS. The hydrophobic interactions of the surfactant tails with P3SBDEAHT are expected to be similar for the three systems. Therefore, the observed differences reflect the influence of electrostatic interactions of the head groups of the surfactants and the ionic side chains of the ZCPE. In fact, for the P3SBDEAHT($C_{12}E_5$) system, no significant interaction was documented in the absorption and fluorescence spectra.

Addition of a zwitterionic surfactant, cocamidopropyl betaine (CAPB), to an aqueous solution of P3SBDEAHT leads to a gradual increase in the fluorescence intensity at concentrations slightly below the *cmc*, accompanied by a blue-shift of ca. 7 nm. The effect on the optical properties is similar to that found for the cationic surfactant. The zwitterionic side chains of the P3SBDEAHT were anticipated to cause strong dipolar interactions with the zwitterionic head group of CAPB and that formation of a zipper-like arrangement similar to that found in the hydrogelation of a zwitterionic poly(fluorene-phenylene) would occur.[55]

However, the distance between the positively and negatively charged centres in the CAPB surfactant is observed to be significantly smaller than that in P3SBDEAHT. Therefore, a zipper-like arrangement is not thought to be likely. The interaction was determined to mainly occur at the periphery of the zwitterionic side chain, similar to the interaction mode postulated for cationic surfactants and in accordance with the similar response of the optical properties during surfactant addition.

MDS was performed to examine aggregates of the tetrameric form of P3SBDEAHT in cationic, anionic, neutral and zwitterionic environments and the results used to draw conclusions on the significance of electrostatics in aggregate formation. In a cationic environment, where the surfactants are DTAC and CTAC, it is apparent that interaction appears predominantly on the outer part of the P3SDBDEAHT:surfactant aggregate, Fig. 4(a) and consequently seems to have little effect on the polymer backbone, this is

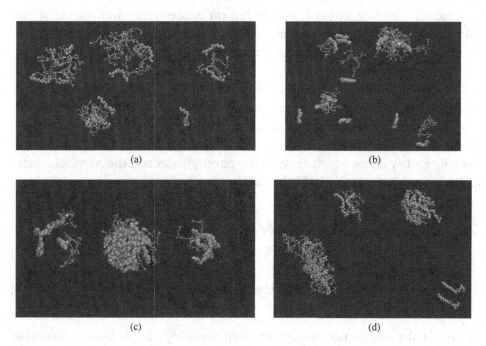

(a)

(b)

(c)

(d)

Figure 4: Simulation cell of P3SBDEAHT and (a) CTAC, (b) SOS, (c) $C_{12}E_5$ and (d) CAPB.

reflected in the minimal changes observed in the photophysical measurements. In the presence of anionic surfactants, in this case SOS, SDS and STS which differ in chain length, it is found that anionic surfactants have a greater effect on the aggregation behaviour of P3SBDEAHT tetramers than cationic surfactants, Fig. 4(b). Interaction between surfactant and P3SBDEAHT is clearly observed. This interaction appears to be more specific where the tail groups of the surfactant become embedded within the aggregate and the head-groups remain exposed to the surrounding solvent.

Upon addition of a zwitterionic surfactant, CAPB, a very clear interaction is observed to take place between the polyelectrolyte and the surfactant through the formation of mixed P3SBDEAHT:CAPB aggregates, Fig. 4(d). Although it is extremely likely that interactions taking place between surfactant and P3SBDEAHT are governed to primarily reduce unfavourable electrostatic interactions, it is possible that the reorientations of the surfactant do not significantly affect the structural rearrangement of P3SBDEAHT and consequently have a minor effect on photophysical behaviour.

Simulations of P3SBDEAHT and neutral $C_{12}E_5$ show that interaction does occur between the polyelectrolyte, Fig. 4(c), and the surfactant but it is possible that the hydrophobic tail group of the surfactant is too long to get embedded into the zwitterion aggregate sufficiently to cause any structural or electronic changes that could be detected by spectroscopy or photochemical techniques. This suggests that the interaction between the nonionic surfactant and the polyelectrolyte takes place predominantly on the outer part of the polyelectrolyte aggregates and has little effect on P3SBDEAHT.

From the study, it was concluded that the nature of the surfactant, anionic, cationic, or zwitterionic has a significant effect on the photophysics of solutions of P3SBDEAHT. The study placed emphasis on the importance of both specific and non-specific interactions within polymer:surfactant assemblies. The charge of the surfactant, and by extension the consequent electrostatic interactions, is therefore a key factor which governs the photophysics and the physiochemical behaviour of zwitterionic P3SBDEAHT.

2.3.   *Molecular dynamics study of self-assembly of aqueous solutions of poly[9,9-bis(4-sulfonylbutoxyphenyl) fluorene-2,7-diyl-2′,2′-bithiophene] (PBS-PF2) in the presence of pentaethylene glycol monododecyl ether ($C_{12}E_5$)*

A molecular dynamics study on self-assembly in aqueous solutions of poly[9,9-bis(4-sulfonylbutoxyphenyl)fluorene-2,7-diyl-2′,2′-bithiophene]

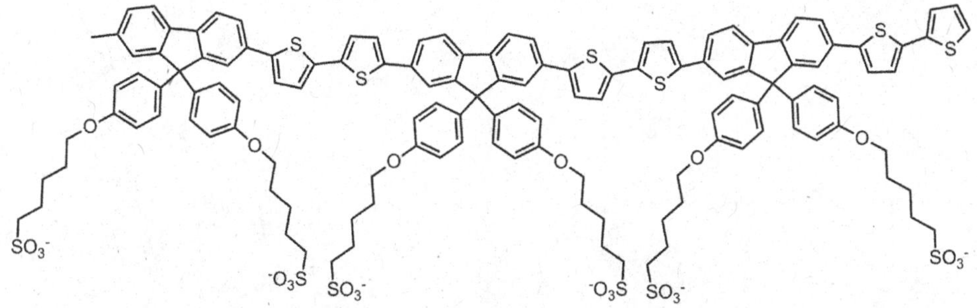

Figure 5: Chemical structure of PBS-PF2T polymer, representative of the trimer used in MDS simulations. (Adapted with permission from Ref. 6.)

(PBS-PF2T), Fig. 5, in the presence of a surfactant pentaethylene glycol monododecyl ether ($C_{12}E_5$), was performed.[61] Simulations were used to examine the interaction between PBS-PF2T and $C_{12}E_5$ and suggest that addition of the surfactant causes the breakup of PBS-PF2T aggregates in solution.

Water soluble conjugated polyelectrolytes (CPEs) are a versatile class of molecules[62,63] where water solubility occurs by virtue of the presence of hydrophilic side chains which comprise, for example, non-charged oligoethyleneoxide,[64,65] or charged alkylammonium groups.[66,67] It is primarily the interaction between CPEs and water which formed the basis of this study. CPEs have applications on charge transport and blocking layers,[68,69] chemosensors[70,71] and biosensors[72,73] but are, however, generally subject to problems surrounding their solubility. Resultantly it is difficult to dissolve CPEs down to the single molecule level and they have a propensity for aggregation without precipitating.[1] Consequently a number of approaches have been implemented to resolve this issue and recover the favourable optical and electronic properties of CPEs. It had been previously demonstrated that the fluorescence of an anionic polyfluorene is enhanced in the presence of a nonionic surfactant pentaethylene glycol.[74] The basis for the strategy of introducing surfactant is centered on the observation that water quenches CPE fluorescence whereas surfactant presence restores it to the level observed in organic solvent by forming an insulating layer between the polymer and the surrounding water.

The study was based on the observation that at sufficient surfactant concentrations binary water-surfacant systems can show a range of lyotropic liquid crystalline (LC) phases.[75]

MDS was used to obtain detailed information on the interactions between PBS-PF2T and $C_{12}E_5$. A simulation cell was used to observe the effect of temperature on a solution of 680 mM $C_{12}E_5$ containing two equivalents of

(a)   (b)

Figure 6:   Simulation cell representation (a) front view; and (b) side view of 680 mM $C_{12}E_5$ with two equivalents of PBS-PF2T at 0°C after 10 ns. (PBS-PF2T is shown in van der Waals representations and solvent is omitted for clarity.) (Reproduced with permission from Ref. 61.)

PBS-PF2T. Simulations were carried out at 0°C, 10°C, 20°C, 45°C, 70°C and 90°C. These simulations were compared to previous simulations in $C_{12}E_4$ and also with a prior single simulation of 10 equivalents of PBS-PF2T which demonstrated aggregation in the absence of $C_{12}E_5$.

Each run was carried out from the same starting arrangement of molecules. Figure 6 shows the simulation after a run of 10 ns at 0°C; it is seen here that, whilst there is a distinctive formation resembling a cylindrical structure, there also seems to be a propensity for one of the PBS-PF2T equivalents to remain partially dissociated in the surrounding solvent. For comparison, Fig. 7 shows a previously studied simulation of 10 equivalents of PBS-PF2T in water where it is clear that, in the absence of any surfactant, PBS-PF2T units interact with each other, resulting in the notable formation of a large PBS-PF2T assembly. It is this type of cluster formation that, it is suggested, leads to a reduction in the fluorescence quantum yield and red-shifted fluorescence maxima. By comparing Fig. 6 and Fig. 7, it was observed that the presence of $C_{12}E_5$ affords a significant inhibition of the aggregation between PBS-PF2T which occurs in a water only environment.

It is found from the dynamics simulations that between 0°C and 10°C this cylindrical phase is maintained where interactions between equivalents of PBS-PF2T are inhibited. At 20°C this phase becomes more dispersed before returning to a cylindrical system at 45°C which persists until 90°C where a diffuse phase is observed again. In the case where $C_{12}E_4$ is used as the surfactant, it is found that at 10°C a spherical micellar phase is formed which became a cylindrical phase at higher temperatures.[76] Additionally, a simulated annealing of PBS-PF2T in $C_{12}E_5$ was performed between 20°C and 10°C to show how the phase changes between each of these temperatures. The annealing began

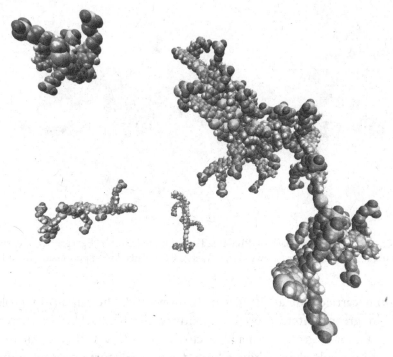

Figure 7:   Simulation cell at 20 ns of ten equivalents of PBS-PF2T in water, PBS-PF2T in van der Waals representation. Reproduced with permission from Ref. 76.

at 20°C which is isotropic in appearance and upon cooling to 10°C the system begins to show the formation of two cylindrical PBS-PF2T:$C_{12}E_5$ assemblies. This finding is consistent with what is observed for individual simulations at these temperatures and provides further validation for the use of MDS as a method to investigate temperature effects on phase formation.

It is apparent, however, in each of the simulations with both $C_{12}E_4$ and $C_{12}E_5$ that there is a specific orientation of the PBS-PF2T units. The polymer is principally incorporated into the surfactant. This incorporation is however rather particular. The charged side chains, explicitly the charged sulfonate termini, of the polymer maintain close contact with the surrounding solvent whereas the backbone remains strongly embedded into the surfactant environment. The side chains accommodate the barrier to inter-unit rotation in the polymer backbone by aligning in different ways to maintain contact with the solvent environment. This is also displayed in the simulation at 20°C and 90°C where phase formation is observed as somewhat more diffuse than that seen at other temperatures. In these simulations the side chain of the PBS-PF2T units appear to interact with the solvent present in what appears to be a reverse micellar section. Figure 8 shows a singular scene from the

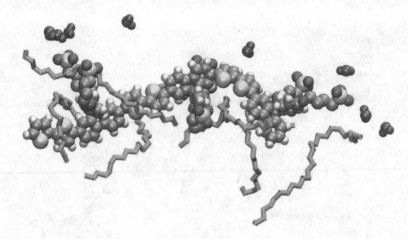

Figure 8: The polymeric part of PBS-PF2T surrounded by $C_{12}E_5$ (green) and sulfonate groups interacting with water (grey van der Waals). Reproduced with permission from Ref. 61.

simulations carried out at 20°C which shows that the surfactant molecules (shown in green) interact predominantly with the backbone whereas the terminal sulfonate groups remain in closer proximity with the surrounding solvent and the side chains show a high degree of distortion to accommodate this sulfonate group-solvent interaction. In order to examine this observation, further analysis was carried out on each of the generated trajectories. This was performed by examining the number of contacts below 0.6 nm between the side chains and the surrounding solvent and comparing these to the number of contacts between the side chains and the surfactants. It was seen that in each case there is a significantly higher number of contacts between PBS-PF2T side chains and the solvent than that which exist between the side chains and the surfactant. This observation strongly supports the hypothesis that the side chains do possess a high degree of hydrophilicity and explain the propensity for the side chains to remain in close contact with the solvent. It is also apparent that the number of contacts between the backbone of PBS-PF2T and the surfactant is also higher than those between the side chains and surfactant which is in agreement with the perceived orientation of the PBS-PF2T units in relation to the surfactant.

This study provides clear evidence for the breakup of PBS-PF2T aggregation upon addition of the nonionic surfactant $C_{12}E_5$, and can be extended to the behaviour with other related alkyloxyethylene surfactants. Aggregate formation is clearly seen to occur between PBS-PF2T equivalents in the absence of surfactant. When this simulation of pure PBS-PF2T is compared to those containing PBS-PF2T in aqueous $C_{12}E_5$/water environments it is seen that this aggregation is inhibited and this is followed by a characteristic phase formation. The formation of rod-like cylinder phases dominates at 0°C, 10°C, 45°C and 70°C. At 20°C and 90°C there is a less defined phase formation, which has some semblance of a reverse micellar phase where there is a circular-type hydrophilic region within which there appears to be an isotropic layer. These findings are in very strong agreement with those which were found in simulations where $C_{12}E_4$ was the surfactant. In the case of $C_{12}E_4$, at 10°C a micellar phase was seen and at 20°C, 45°C, 70°C and 90°C phase formation occurred in the nature of rod-like lamellar structures. In all simulations, whether the surfactant present is $C_{12}E_5$ or $C_{12}E_4$, it is important to observe that the PBS-PF2T species do not interact with each other and resultantly no CPE:CPE (PBS-PF2T:PBS-PF2T) aggregates are formed but instead form distinct polymer:surfactant assemblies. Another very important observation relates to the orientation of the PBS-PF2T units which shows a marked level of specificity and demonstrates an affinity for the backbone of the polyelectrolyte to interact with the surfactant assembly and the side chains to remain exposed to the surrounding solvent environment. These results suggest that it is possible to extend the MD simulations to systems involving alkyloxyethylene surfactants of other alkyl or ethylene oxide chain lengths.

3. What NMR Spectroscopy Tells us About Self-Assembly

Multinuclear NMR spectroscopy has proved to be one of the most versatile of the various spectroscopic methods used for the characterization of structural interactions involved in complexation reactions and in addition the interaction of surfactants, conjugated polymers and polyelectrolytes with metal ion complexes. In complexation reactions, broadening or coordination induced shifts of the ^1H and ^{13}C signals of the ligand in the presence of the metal ions, when compared with those of the free ligand, can provide clear indications of ligand to metal coordination sites, which together with metal ion NMR, suggest valuable structural information, including the type of metal center present in complexes, as widely exemplified in our previous work on the complexation of metal ions with relevant ligands.[77–81] Furthermore, ^1H and ^{13}C NMR spectroscopy is also suitable for understanding the interaction of surfactants, and polyelectrolytes with metal ion complexes and conjugated polymers, as we subsequently describe.

Since the first observation of efficient electroluminescence from aluminium(III) 8-hydroxyquinolate,[82] this class of complexes has been widely studied for opto-electronic applications. Our earlier work on complexes of 8-hydroxyquinoline-5-sulfonate (8-HQS) with metal ions[83–88] show that these form in a pH dependent process, with a maximum stoichiometry of 1:3 (metal:ligand) with Al(III)[83] and Ga(III),[84] and 1:2 with Zn(II),[85] Cd(II),[86] Hg(II),[86] VO_2^+,[87] MoO_2^{2+} and WO_2^{2+}.[88] As with 8-hydroxyquinoline, the free 8-HQS ligand is weakly luminescent, possibly because of the formation of a nonfluorescent phototautomer by fast excited-state intramolecular proton transfer.[89] This is inhibited upon complexation; the lowest excited state is normally localized on the ligand,[90] which leads to dramatic increases in fluorescence intensity on complexation, important for OLEDs and metal ion sensing.[91] With V(V)[87] and Mo(VI),[88] only relatively weak fluorescence is observed due to the presence of an alternative nonradiative decay pathway through low-lying charge transfer states. Other factors which can be important in the photophysics of metal complexes with 8-hydroxyquinoline and 8-HQS include enhanced S_1–T_1 intersystem crossing due to heavy atom effects,[86,92] and nonradiative decay through ligand or solvent exchange.[84,86] This latter nonradiative decay route can be suppressed, and fluorescence yields increased by incorporating metal 8-hydroxyquinolates into solid matrices, such as Amberlite CG-400 ion exchange resins,[93] or adding cationic surfactants, such as cetyltrimethylammonium bromide (CTAB). These processes all involve self-assembly.

We have reported some selected experiments designed to obtain in-depth insights into the origin of this fluorescence enhancement, and also to produce simple formulations which can be used, through self-assembly, to prepare solvent processible luminescent metal hydroxyquinolate derivatives starting from aqueous solutions of commercially available metal salt, 8-hydroxyquinoline-5-sulfonic acid and the surfactants DTAB and CTAB (Scheme 1).[94] The surfactants were chosen as their critical micelle concentrations (*cmc's*) are well separated, which allows us to check the relative importance of electrostatic interactions and micellization (*cmc*: DTAB, 1.56×10^{-2} M; CTAB, 9.2×10^{-4} M).[94] The Al(III) and Zn(II) complexes are chosen as representative

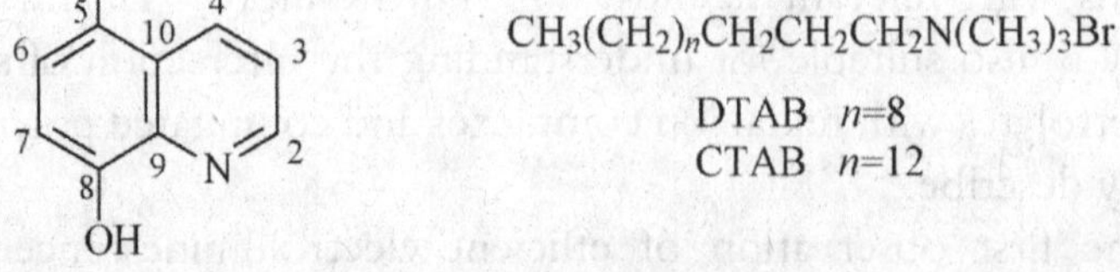

$CH_3(CH_2)_nCH_2CH_2CH_2N(CH_3)_3Br$

DTAB *n*=8
CTAB *n*=12

Scheme 1:   Chemical structures of 8-HQS and surfactants used in this study. For 8-HQS and surfactants, the numbering used in the NMR studies is presented. (Reproduced from Ref. 94 with permission from the Royal Society of Chemistry.)

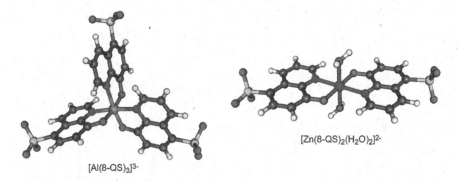

[Al(8-QS)₃]³⁻ [Zn(8-QS)₂(H₂O)₂]²⁻

Scheme 2: Structures of the complex anions based on DFT calculations in Refs. 83 and 85. (Reproduced from Ref. 94 with permission from the Royal Society of Chemistry.)

systems likely to be of interest for electronic applications. Under stoichiometric conditions, the complex formed between 8-HQS and Al(III) in the pH range of 2–6 has a maximum metal: ligand ratio of 1:3, and, as with Alq₃, is the *mer* stereoisomer (Scheme 2).[83] At lower ligand concentrations 1:2 and 1:1 complexes are formed. In contrast, the only significant complex between Zn(II) and 8-HQS is the 1:2 (metal:ligand) species, which forms in the pH range of 6–8, and which has a square bipyramidal geometry, with the two water molecules *trans* to each other, with the remaining coordination positions occupied by the two oxygen and two nitrogen donor atoms of the 8-hydroxyquinoline-5-sulfonate ligands (Scheme 2). We have shown details of the interaction of these two complex ions in aqueous solutions with DTAB, using ^{1}H and ^{13}C NMR spectroscopy (Figs. 9 and 10).

The ^{1}H NMR spectra in the aromatic region show that the 1:3 Al(III)/8-HQS complex is not affected upon addition of the surfactants. In addition, there are no significant differences in the spectra in the presence of DTAB or CTAB, although the concentrations that we studied are below the *cmc* of DTAB and above that of CTAB. This indicates that it is the interaction between the cationic surfactant and the anionic metal complex that is important rather than micellization. There is, however, a slight broadening of the signals in the aromatic region, possibly due to the effect of association with surfactants on the solution dynamics. In contrast, there are differences in the alkyl chain region of the spectra, including the presence of new resonances at lower frequencies for the set of protons labelled as H-5′ suggesting changes in the surfactant conformation on binding to the metal complex.

^{1}H and ^{13}C NMR spectra of the solutions of Zn(II):8-HQS 5:10 mmol dm⁻³ in the presence of surfactants have been obtained as a function of the concentration of the DTAB added. Figures 11 and 12 show the ^{1}H and the

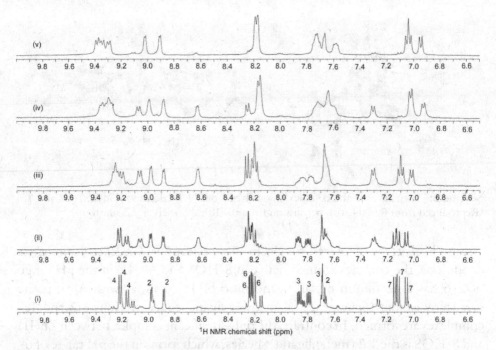

Figure 9: Expansion 6.5–10 ppm of the ^{1}H NMR spectra of D_2O solutions of (i) Al^{3+}/HQS 2.5:7.5 mmol dm^{-3}, pH 5.7, (ii) Al^{3+}/8-HQS/DTAB 2.5:7.5:1.0 mmol dm^{-3}, (iii) Al^{3+}/8-HQS/DTAB 2.5:7.5:5.0 mmol dm^{-3}, (iv) Al^{3+}/8-HQS/DTAB 2.5:7.5:10 mmol dm^{-3} and (v) Al^{3+}/8-HQS/DTAB 2.5:7.5:20 mmol dm^{-3}, temp. 298 K. (Reproduced from Ref. 94 with permission from the Royal Society of Chemistry.)

^{13}C NMR spectra, respectively, for the aromatic region of solutions of Zn(II):8-HQS 5:10 mmol dm^{-3} upon increasing concentration of DTAB.

The ^{1}H NMR spectra in the aromatic region show that the 1:2 Zn(II):8-HQS complexes appear to be unaffected upon addition of small quantities of surfactants. However, upon addition of larger quantities, new broad resonances appear, indicating that the complex could be affected under these conditions, possibly with the ligands undergoing exchange. The ^{13}C NMR spectra shown in Fig. 12, compared with that of the free complex[85] is consistent with this observation, and shows the presence of free 8-HQS.

Another selected example is related with the interaction of the surfactactant n-dodecyl pentaoxyethylene glycol ether ($C_{12}E_5$) with cationic conjugated polyelectrolyte poly{9,9-bis[6-N,N,N-trimethylammonium)hexyl]fluorene-co-1,4-phenylene} (ADS 181, PFP-NR3, Scheme 3).[95]

Proton NMR spectra were recorded in D_2O solutions of $C_{12}E_5$ alone (10 mmol dm^{-3}) and in the presence of various concentrations of ADS 181, and

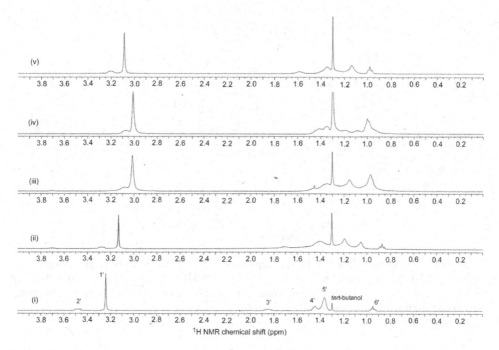

Figure 10: Expansion 0–4 ppm of the ^{1}H NMR spectra of D$_2$O solutions of (i) DTAB 0.1 mol dm^{-3}, (ii) Al^{3+}/8-HQS/DTAB 2.5:7.5:1.0 mmol dm^{-3}, (iii) Al^{3+}/8-HQS/DTAB 2.5:7.5:5.0 mmol dm^{-3}, (v) Al^{3+}/8-HQS/DTAB 2.5:7.5:10 mmol dm^{-3} and (vi) Al^{3+}/8-HQS/DTAB 2.5:7.5:20 mmol dm^{-3}, temp. 298 K. (Reproduced from Ref. 94 with permission from the Royal Society of Chemistry.)

data are given in Fig. 13. Assignment of the peaks was based on literature data.[96]

In the presence of the polymer there were no significant changes in the chemical shifts of the surfactant protons, but changes were observed in the linewidth (given as the peak width at half-height, ($\Delta\nu_{1/2}$), as can be seen in Fig. 13. The peak due to the hydroxyl proton f was not observed due to exchange with the solvent. A well resolved triplet was observed for the methyl group, which was unaffected by the presence of the polymer. However, changes were observed in the widths of the signals attributed to the methylene groups in the presence of polymer. These show up most clearly in the signal attributed to the protons c, which is unaffected by neighbouring groups on the surfactant chain. In the presence of ADS 181, an increase in the NMR linewidth of about 50% was observed with this signal, strongly suggesting a more restricted motion of the alkyl chain of the surfactant in the presence of the polymer due to some kind of polymer–surfactant complexation. This is in excellent agreement with

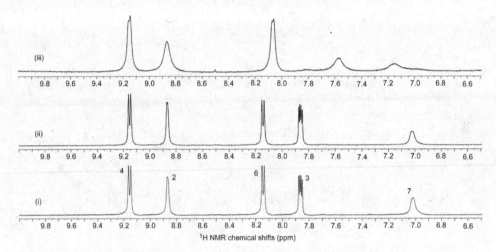

Figure 11: Expansion of the 6.5–10 ppm region of the ^{1}H NMR spectra of D_2O solutions of (i) Zn^{2+}/8-HQS 5:10 mmol dm^{-3}, pH 6.1 (ii) Zn^{2+}/8-HQS/DTAB 5:10:5.0 mmol dm^{-3}, (iii) Zn^{2+}/8-HQS/DTAB 5:10:20 mmol dm^{-3}, temp. 298 K. (Reproduced from Ref. 94 with permission from the Royal Society of Chemistry.)

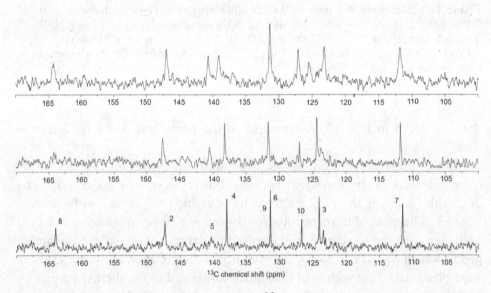

Figure 12: Expansion 100–170 ppm of the ^{13}C NMR spectra of D_2O solutions of (i) Zn^{2+}/8-HQS 5:10 mmol dm^{-3}, pH 6.1 (ii) Zn^{2+}/8-HQS/DTAB 5:10:5.0 mmol dm^{-3}, (iii) Zn^{2+}/8-HQS/DTAB 5:10:20 mmol dm^{-3}, temp. 298 K. (Reproduced from Ref. 94 with permission from the Royal Society of Chemistry).

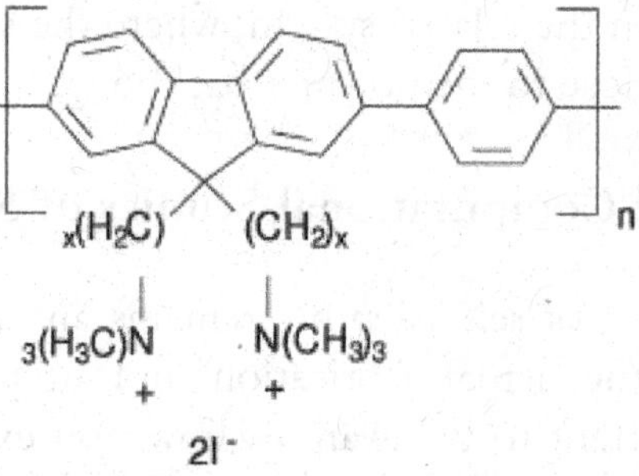

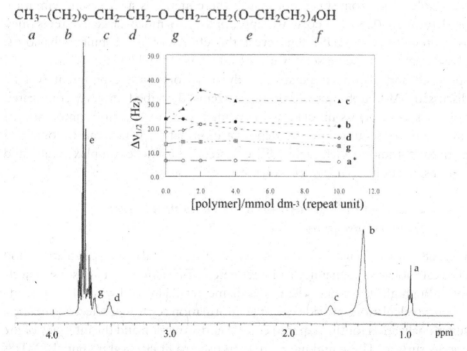

ADS 181   x = 3
PFP-NR3   x = 6

Scheme 3: Structure of the polymer ADS 181. (Reproduced with permission from Ref. 95.)

$CH_3-(CH_2)_9-CH_2-CH_2-O-CH_2-CH_2(O-CH_2CH_2)_4OH$

a   *b*   *c*   *d*   *g*   *e*   *f*

Figure 13: ^1H NMR spectrum with linewidths (inset) observed in D_2O solutions of $C_{12}E_5$ alone (10 mmol dm^{-3}) and in the presence of various concentrations of ADS 181. Note that for the CH_3 protons the coupling constant $J_{CH3}-CH_2$ is used instead of the linewidth. (Reproduced with permission from Ref. 95.)

the results of MDSs on the related system, where the surfactant hydrocarbon chain is seen to bind close to the polymer backbone.

4.   Modelling and Computational Studies of Self-Assembly

Whilst the examination of self-assembly remains an experimental field, ultimately resulting in the actual fabrication and utilization of these super assemblies, it is important to be aware of how non-experimental techniques can lead to detailed insights into the structures and interactions with these systems. Computational techniques can be of great value when it comes to understanding self-assembly. The application of a quantum mechanical or molecular mechanical approach can reveal information on an atomistic level and elucidate the effects individual atoms, bonds, distances and energies can have in the formation of nanostructures. Computational techniques can cover the whole range from small molecules, where methods such as semi-empirical or density functional theory are appropriate, to much larger systems such as proteins where MDSs can reveal the effects of more bulk behaviours. Here, examples of self-assembly examined using a purely quantum mechanical approach and also a cooperative study of theory with experiment will be discussed. Although not all of the systems described involve conjugated polymers, they act as models, as demanded from the cost of computational time, and do demonstrate the depth of information which can be obtained from such studies, which can readily be extended to more complex conjugated systems, including conjugated polymers.

4.1.   *Prediction of ultra-high aspect ratio nanowires from self-assembly*

In 2008 Wu *et al.*[97] used a combination of *ab initio* total energy calculations and classical molecular dynamics to investigate self-assembly of nanoscale objects into ultra high aspect ratio (length vs diameter) chains and wires. In this study *ab initio* calculations provide essential information regarding selective chemical functionalization for the required end-to-end attraction and the interplay of the energy surface. These findings were then used to fit classical potentials. MDSs were carried out to make deductions on the dynamical properties of assembly as functions of synthesis conditions such as solvent, chemical functionalization, temperature and concentration.

The ability to control the synthesis of ultra-high-aspect ratio nanostructures is of extreme importance in next generation devices based on new electronic, optical and mechanical properties at such a size scale.[98–103] The majority of the techniques used in the fabrication of nanowires can be separated into two categories, top down and bottom up methods.[104-114] Top down methods use photo- or electron-beam lithography and dry plasma or wet chemical etching.[107-109] The bottom up approach employs direct chemical synthesis from molecular precursors.[113,114] Although significant progress has been made in nanostructure synthesis, efficient fabrication of nanomaterials with large aspect ratios, (diameters smaller than 50 nm and lengths of μm scale) still presents significant difficulties. Additionally, the ability to manufacture accurate devices using nanoscale building blocks requires the manipulation of nanomaterials into functional and ordered forms and due to their very small size this remains a fundamental challenge.

A promising method for designing and controlling the bottom up approach is self-assembly[9,115] from nanoscale building blocks such as nanospheres,[116] nanorods,[117] nanocubes[118] and nanoplates.[119] This approach has begun to attract a great deal of attention from experimental[120-130] and theoretical researchers.[131-138] Self-assembly has the attractive potential advantage of being cheap, simple and rapid in addition to the electronic and optical properties of the materials produced being comparable and in some ways superior to those obtained by more exhaustive methods.

This work reported a theoretical study of the functionalisation of Si nanorods with small organic molecules as their end surfaces and the corresponding self-assembly under various conditions. The feasibility was demonstrated by using self-assembly to form ultra-high-aspect ratio chains and wires out of these building blocks which are attracted to one another via hydrogen bonds.

A combination of *ab initio* quantum chemical approaches and classical molecular dynamics was used and the parameters in the classical potential model fitted to *ab intio* results. End-to-end attraction is adjusted by controlling the pattern of the rod end chemical functionalization which was found to have a major impact on alignment. *Ab intio* calculations were carried out within a density functional theory approach in conjunction with MDSs. The hydrogen bond is well suited for self-assembly as it is both selective and directional with typical strength greater than van der Waals but less than most ionic or metallic bonds. In this work a simple carboxyl group −COOH (group A) and methylamine group −CH$_2$-NH$_2$ (group B) were used to promote the hydrogen bonding to direct self assembly. Resultantly, the Si nanorod end surfaces are functionalized with either A or B groups.

Test calculations revealed that −COOH pairs (AA) were stronger than −COOH NH$_2$−CH$_2$− (AB) pairs and that both AA and AB pairs were significantly stronger than BB as OH–N and OH–H bonds are stronger than NH–N ones.

Beginning with short Si rods the H-termination at rod ends were replaced with functional groups A or B. Pairs of functional rods were then brought into contact with eachother in a variety of configurations and DFT structural optimization was performed. As expected end-to-end attraction was found to be much stronger than end-to-side.

The goal for the study was the assembly of high aspect-ratio nanowires with a good alignment and it was found that if all H atoms on rod ends were functionalized the outermost (radial) functional groups would attract other rods in a way that would work against parallel alignment. To avoid this only the centre regions of the end surfaces were functionalized whilst the outer edge remained unfunctionalized, Fig. 14. It was found that although partial functionalization leads to improved chain alignment, long chains can also form with full functionalization.

Investigations of solvent effects revealed that competition between solvent molecules and functional groups to bond rods together makes polar solvents inefficient for self-assembly whereas non-polar solvent CCl_4 and toluene do not block functional groups from bonding together.

At higher temperatures assembly occurs more rapidly and the collision rate constant among building blocks is independent of their concentration if the average separation of rods or chains is much longer than their length, Fig. 15(a).

The *ab initio* study of the formed wires showed both the highest occupied molecular orbital (HOMO) and lowest unoccupied molecular orbital (LUMO) of the −COOH groups fall below and above the nanorod gap.

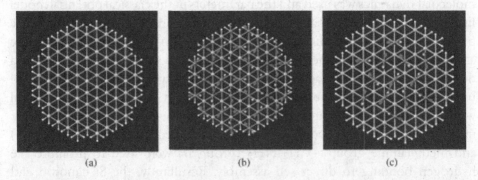

(a)   (b)   (c)

Figure 14:   Effects of surface patterning on self-assembly. The panels show configurations of rod ends with (a) terminating H atoms, (b) replacing all H atoms with functional group A and (c) partially replacing H atoms with functional group A. (Adapted with permission from Ref. 97. Copyright 2008 American Chemical Society.)

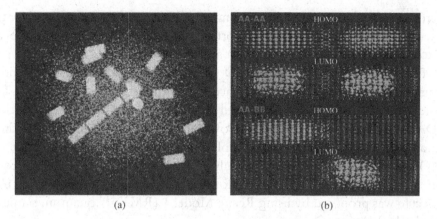

Figure 15: (a) Snapshot of a simulated self-assembly process and (b) isosurfaces (green) of the charge density of the HOMO and LUMO orbitals for the AA–AA chain above and the AA–BB chain below. (Adapted with permission from Ref. 27. Copyright 2008 American Chemical Society.)

For $-CH_2-NH_2$, the LUMO is also well above the energy gap whereas the HOMO is very close to the top of the valence states. Resultantly the AA-type functionalization may be more favourable than BB-type for certain electronic applications.

In terms of HOMO and LUMO localization in the AA–AA chain the HOMO and LUMO are located on every building block. In the AA–BB chains the HOMO and the LUMO are separated to the AA-side and BB-side respectively therefore forming a type II junction between functionalized rods due to hydrogen assisted electron transfer, Fig. 15(b).

The calculations presented point towards a possible novel and efficient route for synthesizing these ultra- high-aspect ratio nanowires. This type of functionalization could also be of great benefit in the fabrication of and self-assembly of other nanostructures such as carbon nano-tubes and nanosheets.

4.2. Self-assembly of acridine orange into H-aggregates at the air/water interface: Tuning of orientation of headgroup

In 2011, Jiménez-Millán *et al.*[139] published their findings on self-assembly of Acridine Orange into H-aggregates at the air/water interface.

In this study, the surface active derivative of the organic dye Acridine Orange (*N*-10-dodecyl-acridine orange (DAO)) was included in mixed Langmuir monolayers with stearic acid (SA). For a stable mixed monolayer the maximum relative content on DAO is a molar ratio of $X_{DAO} = 0.5$. Brewster angle microscopy (BAM) revealed a high homogeneity at the micrometer level for the mixed monolayer in equimolar proportion, $X_{DAO} = 0.5$, whereas the appearance of domains occurs at lower content of DAO i.e., $X_{DAO} = 0.2$ and 0.1.

The aggregation of the DAO head-group leads to well defined H-aggregates at the air/water interface for the mixed monolayer with a low content of DAO. However it was found that for the mixed monolayers enriched in DAO e.g. $X_{DAO} = 0.5$ molecular crowding predicts the formation of defined supramolecular structures. Whilst molecular organisation and tilting of the DAO head-group is quantitatively analysed by UV-vis reflection spectroscopy, a molecular mechanism for the conformational rearrangement of the DAO molecule was proposed by using Recife Model 1 (RM1)[140] quantum semiempirical calculations.

As functional building blocks for supramolecular assemblies, photoactive molecules are of maximum relevance.

Organic dyes have been used for studies involving Förster resonance energy transfer[141–143] in biophysics offering important new insights in the conformations and mechanisms of biomolecules. Organic dyes also play an important role in organic electronics,[144–147] e.g. organic molecular antennas, mimicking the natural systems which are designed for maximum capability of converting solar light into electrical energy. As a consequence both the photophysical features and molecular arrangement are key parameters for the overall efficiency of photovoltaic cells.[148,149] Therefore controlling and tuning the molecular arrangement of organic dyes is highly desirable for supramolecular design.

H-aggregates of a dye occur in the case of parallel arrangement of the dye units (face to face), Fig. 16, which leads to a blue shift in the absorbance

Interfacial self-assembly

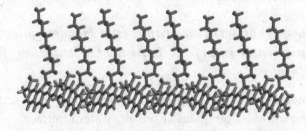

H-aggregates

Figure 16:   Schematic of the parrallel arrangement of dye units face to face. (Adapted with permission from Ref. 139. Copyright 2011 American Chemical Society.)

band.[150] Controlling formation of H-aggregates of organic molecules is a current challenge in organic synthesis as the H-aggregates have proved to enhance stable morphologies in thin films leading to reproducible features in organic thin film transistors.[151]

The Langmuir technique is of great use in this field as it allows for the preparation of a monomolecular layer at the air/water interface. As a two-dimensional (2D) Langmuir monolayer is molecularly well defined as well as easily prepared. The Langmuir technique also offers accurate control of the molecular area and the application of a certain surface pressure.[152,153] Therefore, the design and study of Langmuir monolayers containing organic dyes are interesting options for gaining insights into supramolecular assemblies.

The work presents an amphiphilic derivative of acridine orange (DAO). The DAO molecule is formed by an acridine polar group and a single hydrocarbon chain covalently attached to the acridine moiety. To enhance the stability of the DAO presence at the interface stearic acid (SA) composed of a single hydrocarbon chain and a small head-group was included in the monolayer, leading to the study of mixed Langmuir monolayer with different molar ratios.

Investigations of the surface pressure-molecular area (π-A) isotherms were performed to determine thermodynamic information about the DAO:SA mixed monolayers. It was found that mixed monolayers with a molar excess of DAO ($X_{DAO} > 0.5$) were not stable at the air/water interface and a significant solubilization of DAO into the subphase was observed. Mixed monolayer with $X_{DAO} \geq 0.5$ were not soluble in the subphase at the air/water interface. Moreover, from performing two subsequent complete π-A isotherms it was found that the amount of DAO at the interface is kept constant and additionally the molecular arrangement of the DAO molecule is reversibly achieved. The π-A isotherms of mixed monolayers showed a liquid expanded (LE) to a liquid condensed (LC) phase transition at 12–15 mN/m surface pressure where the surface pressure for the phase transition depends on the molecular ratio of the DAO in the mixed monolayer. There were interactions between DAO and SA in the mixed monolayer and were therefore considered to be miscible to a reasonable extent.

UV-Vis spectral analysis from the mixed monolayer showed two main phenomena involving the DAO head-group, firstly the formation of H-aggregates, indicated by a blue shift of UV-vis bands and secondly a change in the tilting of the main dipole of the DAO head-group was observed with applied surface pressure. It was also revealed that the formation of H-aggregates is enhanced for mixed monolayers with a lower content of DAO. Although in such cases the absolute content of DAO molecules is reduced for these mixed monolayers it was concluded that the absence of molecular crowding at the interface favours the formation of linear and well-defined H-aggregates.

Tilting of the head-group was tuned by applying a certain surface pressure to a given monolayer. The tilting of the angle formed between the main axis of the DAO head-group and the z axis, θ, ranged from $\theta = 90°$ (low X_{DAO} and low applied surface pressure) to $\theta = 55°$ (large applied surface pressure). The interval of θ that a given mixed monolayer can achieve was found to be controlled by using a certain molar ratio of DAO. Although the experimental techniques described above give accurate information on the arrangement of the DAO head-groups at the air/water interface they do not provide mechanistic insights.

A significant tilting of DAO head-groups was observed for all mixed monolayers and whilst this was expected for mixed monolayers with $X_{DAO} = 0.5$ given the larger content of DAO at the interface, for mixed monolayers with low content of DAO, $X_{DAO} = 0.2$ no tilting might be expected as DAO molecules are likely to be distant from one another. However, experimental studies provided evidence of tilting and aggregation for even reduced values of DAO suggesting that a driving force for tilting other than the effect of a large enough surface concentration of DAO at the air/water interface exists.

Using a semiempirical quantum chemical approach with the RM1 method, different conformations of the DAO molecule were analysed to determine the most stable arrangement. Using this conformation of the DAO molecule at the air/water interfact the formation of the simplest aggregate — a dimer — between the DAO molecules was checked. Although calculations were performed in vacuo they offered valuable insights into rearrangement at a molecular level.

Figure 17 shows the optimized molecular structure of DAO as well as the atom numbering for those atoms involved in conformational change. The main rotation concerning the tilting of the DAO headgroup was assumed to take place around the bond between C(4) and C(5) atoms, i.e. the dihedral bond C(3)-C(4)-C(5)-C(6) should rearrange with compression of the monolayer. After calculating the values of the total electronic energy for DAO as a function of dihedral angle C(3–6) angle values of 70° and 180° were found to be energy minima. As 180° corresponds to an all trans arrangement this was determined to be unlikely to form at the air/water interface and therefore a C(3–6) value of 70° was determined as the most likely to occur for DAO at the air/water interface, Fig. 17.

This angle was rotated from 180° to 70° and, by assuming a vertical orientation of the hydrophobic chain with respect to the air/water interface a value of approximately 50 Å^2 per DAO molecule was calculated. Additionally

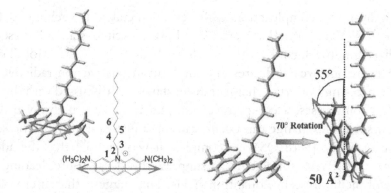

Figure 17: Molecular scheme for the most stable conformation of DAO (left). The red arrow indicates the main transition dipole of the DAO headgroup. Rotation of the dihedral plane C(3–6) from 180° to 70° (right). Adapted with permission from Ref. 139. Copyright 2011 American Chemical Society.

the angle found by the transition dipole moment of the DAO head-group and z-axis was calculated as approximately 55°. These values were in excellent agreement with experimental results obtained for the molecular layer and the tilting of the head-group for the mixed monolayer $X_{DAO} = 0.5$ and the surface pressure. Therefore it was proposed that the molecular mechanism for the DAO rearrangement at the interface is the rotation around the C(3–6) dihedral angle as a function of applied surface pressure.

It was concluded that the spontaneous formation of H-aggregates at the mixed monolayer DAO:SA is driven by the tendency of the DAO head-groups to self-assemble. The tilting of the head-group as well as the degree of aggregation can be accurately tuned by adjusting the molar ratio of DAO and the applied surface pressure. Additionally, the process of aggregate formation is found to be reversible.

4.3. Interaction of 8-hydroxyquinoline-5-sulfonate with metal ions and surfactants

The study of self-assembled systems between metal complexes and cationic alkylammonium surfactants can be used to model the interactions between metal complexes and conjugated polyelectrolytes, since the surfactants can be chosen to have the same structure as the side chains of the conjugated polyelectrolytes. The self-assembly between the $[Al(8\text{-}QS)_3]^{3-}$ complex[83] (8-HQS = 8-hydroxyquinoline-5-sulfonate) and the cationic cetyltrimethy-lammonium bromide (CTAB) surfactant is driven predominantly through electrostatic interactions between systems with oppositely charged groups. The free 8-HQS ligand is weakly luminescent, however its fluorescence increases dramatically upon complexation with certain metal ions such as Al(III),[83] Ga(III),[84] W(VI),[88] Zn(II),[85] etc. The lowest excited state in these cases is localized on the ligand (with metals such as V(V)[87] and Mo(VI)[88] the complexes are only weakly fluorescent due to an alternative non-radiative decay pathway involving low-lying charge transfer states). As discussed with the NMR results, in some cases, a competitive non-radiative decay pathway involving ligand or solvent exchange can occur, causing a decrease in the fluorescence. In the case of the $[Al(8-QS)_3]^{3-}$ complex, it was found[94] that the addition of the cationic CTAB surfactant can suppress this decay route, leading to an increase of the fluorescence quantum yields. This suggests that intermolecular interactions are established between the two systems and that self-assembled complex-surfactant systems are formed. To confirm this association, density functional theory (DFT) calculations were carried out for the $[Al(8-QS)_3]^{3-}$ complex and for a model of the assembled system using the B3LYP[154–157] functional and the 6–31G(d) basis set with the GAMESS-US[158] program. The model of the assembled system considers the metal complex structure surrounded by three N,N,N-trimethyl-N-propylammonium (TMPA) cations, each one in the proximity of one of the sulfonate groups of the 8-HQS ligands (the use of TMPA cations in the calculation instead of the CTAB cations allows to reduce the computational time without modifying the relevant characteristics of CTAB). Density functional theory (DFT) methods allow the treatment of electron correlation, which is very important for large molecules and systems with heavy atoms, at a much lower computational cost.[159,160] The DFT optimized geometries, both *in vacuo* (Fig. 18(a)) and in a simulated water environment (using a polarizable continuum model, PCM[161,162]) (Fig. 18(b)), indicated the establishment of electrostatic interactions between the sulfonate groups of the complex and the TMPA counterions. The shortest distances between the N atom of the cation and the O atoms of sulfonate in each group are 3.808, 3.821 and 3.754 Å in the geometry obtained *in vacuo*, and 3.639, 3.608 and 3.570 Å in the geometry optimized in the simulated water environment. In the latter geometry, the delocalization of charge in the sulfonate groups is not complete, and the cations interact more strongly with one or two of the oxygen atoms in each sulfonate group, as seen in Fig. 18(b). Table 1 gives some selected structural parameters for the $[Al(8-QS)_3]^{3-}$ complex optimized with PCM, in the presence or absence of TMPA counterions. The bond lengths, angles and charges suffer only minor changes with the presence of the counterions, showing that the complex structure remains intact and the charges around the central aluminium atom are, basically, unaffected by the interaction with the counterions. These results show that a

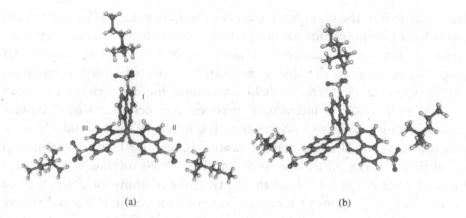

(a)   (b)

Figure 18: (a) B3LYP/6-31G(d,p) optimized geometry *in vacuo* of an aggregate of the meridional $[Al(8\text{-}QS)_3]^{3-}$ complex and three N,N,N-trimethyl-N-propylammonium ions; b) the optimized geometry considering the bulk solvent (water) effects through PCM.[161,162] (Adapted from Ref. 94 with permission from the Royal Society of Chemistry.)

Table 1: Selected bond lengths (Å) and angles (degrees) calculated at the B3LYP/ 6-31G(d,p) level with PCM (water)[a] for the meridional $[Al(8\text{-}QS)_3]^{3-}$ complex in the presence or absence N,N,N-trimethyl-N-propylammonium counterions.

Bond lengths	$[Al(8\text{-}QS)_3]^{3-}$ + counterions	$[Al(8\text{-}QS)_3]^{3-}$	Angles	$[Al(8\text{-}QS)_3]^{3-}$ + counterions	$[Al(8\text{-}QS)_3]^{3-}$
$Al\text{-}O_I$	1.878	1.885	$O_I\text{-}Al\text{-}N_I$	80.84	80.75
$Al\text{-}O_{II}$	1.870	1.870	$O_{II}\text{-}Al\text{-}N_{II}$	83.04	83.74
$Al\text{-}O_{III}$	1.884	1.887	$O_{III}\text{-}Al\text{-}N_{III}$	82.56	83.55
$Al\text{-}N_I$	2.109	2.111	$O_I\text{-}Al\text{-}O_{II}$	93.46	92.70
$Al\text{-}N_{II}$	2.078	2.073	$N_I\text{-}Al\text{-}O_{III}$	89.09	89.95
$Al\text{-}N_{III}$	2.063	2.046	$N_{II}\text{-}Al\text{-}N_{III}$	171.70	173.67

[a]Polarizable Continuum Model.[161,162]

self-assembled system between the $[Al(8\text{-}QS)_3]^{3-}$ complex and the surfactant is stable and is established without disrupting the metal complex.

4.4. *Computational studies on self-assembly in β-phase formation*

Poly(9,9-dioctylfluorene-2,7-diyl) (PF8)[163,164] and some other linear side chain poly(9,9-dioctylfluorenes[165] show very interesting self-assembly behaviour in poor solvents by adopting, under certain temperature and processing conditions, an almost planar conformation (the so called beta-phase). This conformation shows enhanced emission in light emission devices and a red shift in the photoluminescence. The interactions which lead to the formation of planar conformations in conjugated polymers having alkyl side-chains are not yet well understood, however, there is evidence that van der Waals forces between the side chains may be relevant to the stabilization of the beta-phase in PF8. To help understand the aggregation behaviour of these polymers, the interactions between polymer chains of PF8 were studied in toluene and chloroform solutions using NMR spectral and relaxation measurements, small angle neutron scattering and DFT and semiempirical calculations.[166] The NMR resonances of PF8 in both solvents were assigned by computing at the DFT level the NMR chemical shifts for the optimized geometry of the monomer. Chloroform is a good solvent for PF8 and therefore very few polymer-polymer interactions are expected. On the other hand, in a poorer solvent such as toluene, the formation of aggregates is expected. The spin-lattice relaxation times of the ^{13}C and ^{1}H atoms of the alkyl chains of PF8 were measured in concentrated solutions in toluene and chloroform. These suggested stronger polymer-polymer interactions in toluene, involving some overlap or interdigitation of the alkyl chains. This was confirmed using computational methods. For molecules with hundreds or thousands of atoms, the computational cost of *ab initio* or DFT treatments becomes prohibitive. For these systems, semiempirical molecular orbital methods are a valuable tool, representing a good compromise between accuracy and feasibility. These methods involve approximations to simplify the solution of the Hartree-Fock equations, while at the same time possibly also improve their accuracy, by introducing parameterizations to reproduce key experimental quantities.[159] The aggregation behavior of PF8 was simulated using semiempirical PM3 and PM6 quantum chemistry calculations using the MOPAC2007[167] program for a system composed of two PF8 trimers and these calculations allowed to confirm a strong tendency of PF8 to aggregate through interdigitation of the alkylic chains, which was attributed to the establishment of van der Waals attraction between the alkyl chains.[166] Figure 19 shows the PM3 optimized structure for this system.

4.5. *Self-assembly by π-stacking in carbazoles*

Carbazole derivatives bearing phenylisoxazolyl groups have the potential to self-assemble through dipole-dipole and π-π stacking interactions. The self-assembled structure of 3,6-bis(3-(4-decyloxyphenyl)isoxazolyl)-*N*-{4-(3-phenylisoxazolyl)-phenyl} carbazole, Fig. 20)[168] in decalin was analyzed by X-ray diffraction and revealed a stacked structure with a mean distance between

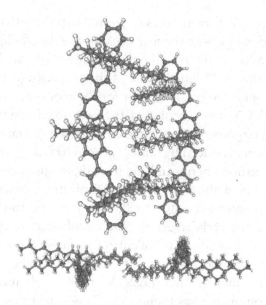

Figure 19: Two perspectives of the PM3 optimized geometry of an aggregate of two F8 trimers in vacuum. (Adapted with permission from Ref. 166. Copyright 2011 American Chemical Society.)

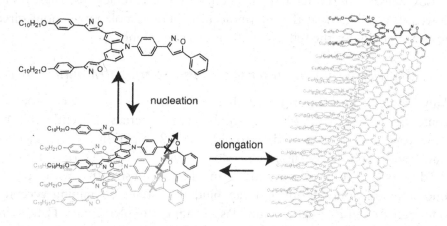

Figure 20: Molecular structure of 3,6-bis(3-(4-decyloxyphenyl)isoxazolyl)-*N*-{4-(3-phenylisoxazolyl)phenyl} carbazole and schematic representation of its cooperative self-assembly. Arrows indicate the dipole moment of isoxazole rings. (Adapted with permission from Ref. 168. Copyright 2011 American Chemical Society.)

the stacked carbazoles consistent with the typical π-π stacking distance. Additionally, the dipoles of the twisted isoxazole rings substituted on the nitrogen of the carbazole unit were found to be aligned in a head-to-tail fashion along the stacking direction and almost perpendicular to the carbazole plane. These structural characteristics suggested that the self-assembling of this compound was driven by both the π-π stacking and dipole-dipole interactions. To understand the cooperativity in the self-assembled system, DFT calculations employing the M06-2X[169] functional and the 6–31G(d,p) basis set with the Gaussian09[170] program were carried out for the monomer, dimer, trimer and tetramer, using the X-ray structure of the assembled system as input. The geometries of the oligomers were extracted from the crystal structure and the decyl side chains were replaced by methyl groups to reduce the computational time. The complexation energy and the total dipole moment were calculated for these geometries without structure optimization. Since each oligomer has two possible conformers, due to flexibility in the torsion between the phenyl group and the carbazole unit, the calculations were carried out for two conformers of each size. The calculated complexation energies were all negative, showing that the assembly is favourable. For the dimer, trimer and tetramer, the dipole moment per molecule (total dipole moment divided by the number of monomers) was found to be larger than that of the monomer and to increase as the size of the oligomer increases. This indicated that the association of monomers into the assembly induces electronic polarization of the isoxazole ring, resulting in stronger dipole-dipole interactions. These results allowed to conclude that the intermolecular dipole-dipole interactions direct the cooperativity for the self-assembly of this compound.

4.6.   *Hydrogen bonding interactions in π -conjugated materials*

Many studies have demonstrated that H-bonding is an efficient tool for promoting the self-assembly of π-conjugated materials. Recently this has been shown for a series of dipyrrolo[2,3-*b*:3',2'-e]pyridine (P2P) electron donors and naphthalenediimide/perylenediimide (NDI/PDI) acceptors.[171] P2P donors have three nitrogen atoms in a fused-ring structure with a proton donor-acceptor-donor motif and can bind to systems containing acceptor-donor-acceptor units, such as diimides, forming complementary H-bonding complexes. The P2P and NDI compounds were mixed in non-polar solvents and this led to change in colour of the solutions and precipitation. The chromism was found to result from a charge-transfer (CT) absorption which emerged upon H-bonding complexation and π-π–stacking of the H-bonded complexes. DFT calculations using the B3LYP[154–157] functional with the 6–31G(d) basis set and the Gaussian09 program[170] were carried out in order to understand the effect of H-bonding and π-stacking on the electronics of the assembled systems. The H-bonding leads to an increase of the HOMO energies of the P2P donors and to stabilization of the LUMO of the NDI acceptors,

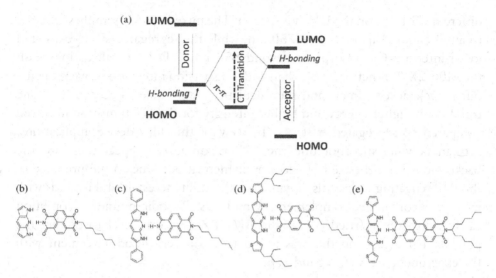

Figure 21: (a) Energy diagram showing the electronic perturbation due to H-bonding and π–π stacking. (b)–(e) Images of some co-assemblies considered in this study. (Adapted from Ref. 171.)

Fig. 21. As a result, in the H-bonded complexes the HOMO-LUMO offset was found to be smaller than that calculated for the free components. The HOMO and LUMO orbitals in the bound system do not overlap and remain localized on the donor and acceptor, respectively. The smaller HOMO-LUMO offset was interpreted as resulting from the net displacement of protons from P2P to NDI. The vertical ionization potential (IP) and the electron affinity (EA) were also calculated for the free and H-bonded systems and the IP values of the P2P donors were found to decrease upon H-bonding while the EA values of the NDI acceptors were found to increase. The ΔEA values were found to correlate with the strength of the H-bond. These results showed that H-bonding can be used to tune the electronic structure in neutral semiconductors. The adiabatic IP and EA were also calculated for the radical cation and the radical anion of the H-bonded assemblies and the adiabatic values were found to show larger changes upon H-bonding compared to the vertical values and this was attributed to enhanced relaxation of charges in the assemblies. The binding energy in the H-bonded charged complexes is also larger compared to the neutral complexes. Cocrystals of the H-bonded P2P donors and NDI acceptors were prepared and their X-ray structures revealed two types of packing motifs: segregated stacks and mixed stacks. The segregated stacks occurred for the cases where the steric interactions did not disfavour self-stacking and the donor-acceptor interactions were weak. The two different types of packing motifs were found to give rise to different CT bands in the UV/Vis spectra. The mixed-stack assemblies gave rise to an intense CT band at 450–750 nm, while the segregated stacks exhibited lower-intensity CT absorption beyond 850 nm. TD-DFT calculations with the M06-2X[169] functional, which includes dispersion interactions, were carried out for π–stacked dimers and the calculations were able to predict the same trend, with higher energy and higher intensity for the CT transition in mixed compared to segregated π-stacks. In view of thin-film device applications, a furan polymer incorporating the P2P π-conjugated repeat unit into the backbone was synthesized by electropolymerization. Due to the presence of the P2P repeating units, this polymer has the ability to establish H-bonds with molecules containing complementary moieties. Periodic boundary condition calculations were carried out at the B3LYP/6–31G(d) level to predict the polymer band gap and this was found to be in very good agreement with the experimental UV/Vis band-gap.

4.7.   *Computational studies of a water soluble poly(aniline)*

Water-soluble materials offer many advantages over materials processed from organic solvents, such as biocompatibility and simple operation and in some cases water can promote self-assembly. The self-assembly of the emeraldine base (EB) state of a water soluble tetra(aniline)-based cationic amphiphile, TANI-NHC(O)C_5H_{10}N(CH$_3$)$_3^+$Br$^-$ (TANI-PTAB), was studied in aqueous solution.[172] The solubility of TANI in water is very low, however, by introducing a cationic headgroup, the derivative produced has improved water-solubility. Aqueous solutions of the EB state of TANI-PTAB were found to show concentration-dependent shifts in UV/Vis absorbance spectra, suggesting the formation of aggregates. DFT calculations were carried out to support the assignments of the experimental UV/Vis spectra and to assist in determining the structures of the aggregates. The geometry of EB TANI-PTAB was optimized using the B3LYP[154–157] functional and the 6–31G(d) basis set with the Gaussian 09 program.[170] A polarizable continuum model (PCM) solvation field[161,173] with water as a solvent was also used. TD-DFT calculations with the CAM-B3LYP[174] functional were then performed for the B3LYP optimized geometry to calculate the excitation energies for the first 20 singlet transitions. The TD-DFT results allowed the assignment of a broad absorbance in the UV/Vis spectrum at 550 nm to the HOMO-LUMO transition of aqueous TANI-PTAB and also indicated that the transition dipole moment (TDM) of the HOMO-LUMO transition is oriented along the long axis of the TANI moiety, Figure 22(a). Concerning the aggregate morphology of EB TANI-PTAB, transmission electron microscopy (TEM) and cryo-TEM

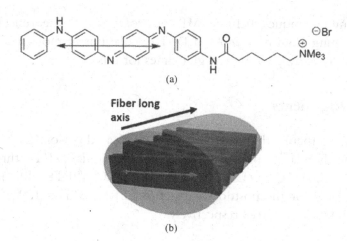

(a)

(b)

Figure 22: (a) TD-DFT calculated orientation of the transition dipole moment of the HOMO-LUMO transition relative to the EB TANI-PTAB molecular structure. (b) Proposed alignment of TANI chromophores within nanofibers, with transition dipole moment (white arrow) overlaid, relative to fiber axis (black arrow). Adapted from Ref. 172.

revealed highly anisotropic nanowires with an individual width of 3 nm, which is comparable to the length calculated for a single TANI-PTAB molecule (2.9 nm). Aligned samples were then studied using linear dichroism (LD) to investigate the molecular structure of the nanowires in solution. LD results indicated a preference for absorbance of light polarized perpendicularly to the direction of the long axis of the nanowires. Taking into account the TD-DFT result that the TDM of the HOMO-LUMO transition is oriented along the long axis of TANI, the LD results suggest that, within the nanowires, TANI-PTAB molecules are arranged normal to the long axis of the nanowires Fig. 22(b). Due to its tunable conductivity, supramolecular morphology and facile aqueous processing this new material was found to have great potential for use in devices and sensors.[172]

5. Conclusions and Outlook

Self-assembly in conjugated polymers, polyelectrolytes and related materials used in organic electronics provides an attractive route to improving solution properties, such as solubility, and form hybrid materials. The understanding of the various interactions involved, such as electrostatic forces, hydrogen bonding, π-stacking, hydrophobic interactions and steric forces, is suggesting routes to the specific design of advanced materials for various electronic and optoelectronic applications. The combination of information from structural studies using techniques such as NMR spectroscopy with theoretical methodologies, ranging from *ab initio* and DFT calculations to molecular dynamics simulations, is starting to suggest guidelines for this.

Acknowledgements

The authors thank the Centro de Química de Coimbra for funding by the Fundação para a Ciência e Tecnologia (FCT) through the programmes UID/QUI/UI0313/2013 and COMPETE. BS and LLGJ thank the FCT for the postdoctoral grants SFRH/BPD/82396/2011 and SFRH/BPD/97026/2013 respectively.

References

1. H. D. Burrows, M. Knaapila, S. M. Fonseca, T. Costa, Chapter 4, Aggregation properties of conjugated polyelectrolytes, Edited by B. Liu , G. Bazan, *Conjugated Polyelectrolytes* ISBN 978-3-527-33143-7 — Wiley-VCH Verlag GmbH & Co. KGaA, (2012) p. 127–167.
2. K. T. Kamtekar, A. P. Monkman, M. R. Bryce, Recent advances in white organic light-emitting materials and devices (WOLEDs). *Adv. Mater.*, **22**, 572–582 (2010).
3. S. Günes, H. Neugebauer, N. S. Sariciftci, Conjugated polymer-based organic solar cells. *Chem. Rev.*, **107**, 1324–1338 (2007).
4. A. C. Grimsdale, K. Leok Chan, R. E. Martin, P. G. Jokisz, A. B. Holmes, Synthesis of light-emitting conjugated polymers for applications in electroluminescent devices. *Chem. Rev.*, **109**, 897–1091 (2009).
5. H. D. Burrows, S. M. Fonseca, F. B. Dias, J. S. de Melo, A. P. Monkman, U. Scherf, S. Pradhan, Singlet excitation energy harvesting and triplet emission in the self-assembled system poly{1,4-phenylene-[9,9-bis(4-phenoxy-butylsulfonate)]fluorene-2,7-diyl} copolymer/tris(bipyridyl)ruthenium(II)in aqueous solution. *Adv. Mater.*, **21**, 1155–1159 (2009).
6. S. M. Pinto, H. D. Burrows, M. M. Pereira, S. M. Fonseca, F. B. Dias, R. Mallavia, M. J. Tapia, Singlet–singlet energy transfer in self-assembled systems of the cationic poly{9,9-bis[6-N,N,N-trimethylammonium)hexyl]fluorene-co-1,4-phenylene} with oppositely charged porphyrins. *J. Phys. Chem. B*, **113**, 16093–16100 (2009).
7. R. C. Huber, A. S. Ferreira, R. Thompson, D. Kilbride, N. S. Knutson, L. S. Devi, D. B. Toso, J. R. Challa, Z. H. Zhou, Y. Rubin, B. J. Schwartz, S. H. Tolbert, Long-lived photoinduced polaron formation in conjugated polyelectrolyte-fullerene assemblies. *Science*, **348**, 1340–1343 (2015).
8. J.-M. Lehn. *Supramolecular Chemistry*, ISBN 978-3-527-29312-4 — VCH Verlagsgesellschaft mbH (1995).
9. G. M. Whitesides, B. Grzybowski, Self-assembly at all scales. *Science*, **295**, 2418–2421 (2002).

10. G. M. Whitesides, M. Boncheva, Beyond molecules: Self-assembly of mesoscopic and macroscopic components. *Proc. Natl. Acad. Sci.*, **99**, 4769–4774 (2002).

11. K. J. M. Bishop, C. E. Wilmer, S. Soh, B. A. Grzybowski, Nanoscale forces and their uses in self-assembly. *Small*, **5**, 1600–1630 (2009).

12. R. C. Evans, Harnessing self-assembly strategies for the rational design of conjugated polymer based materials. *J. Mater. Chem. C*, **1**, 4190–4200 (2013).

13. H. D. Burrows, A. J. M. Valente, T. Costa, B. Stewart, M. J. Tapia, U. Scherf, What conjugated polyelectrolytes tell us about aggregation in polyelectrolyte/surfactant systems. *J. Mol. Liq.*, **210**, 82–99 (2015).

14. T. Van der Boom, R. T. Hayes, Y. Zhao, P. J. Bushard, E. A. Weiss, M. R. Wasielewski, Charge transport in photofunctional nanoparticles self-assembled from zinc 5,10,15,20-tetrakis(perylenediimide)porphyrin building blocks. *J. Am. Chem. Soc.*, **124**, 9582–9590 (2002).

15. M. R. Wasielewski, Photoinduced electron transfer in supramolecular systems for artificial photosynthesis. *Chem. Rev.*, **92**, 435–461 (1992).

16. D. Gust, T. A. Moore, A. L. Moore, Mimicking photosynthetic solar energy transduction. *Acc. Chem. Res.*, **34**, 40–48 (2001).

17. R. W. Wagner, J. S. Lindsey, J. Seth, V. Palaniappan, D. F. Bocian, Molecular optoelectronic gates. *J. Am. Chem. Soc.*, **118**, 3996–3997 (1996).

18. A. P. Silva, H. Q. N. Gunaratne, T. Gunnlaugsson, A. J. M. Huxley, C. P. McCoy, J. T. Rademacher, T. E. Rice, Signaling recognition events with fluorescent sensors and switches. *Chem. Rev.*, **97**, 1515–1566 (1997).

19. I. Willner, B. Willner, Layered molecular optoelectronic assemblies. *J. Mater. Chem.*, **8**, 2543–2556 (1998).

20. J. M. Tour, M. Kozaki, J. M. Seminario, Molecular scale electronics: A synthetic/computational approach to digital computing. *J. Am. Chem. Soc.*, **120**, 8486–8493 (1998).

21. C. P. A. Collier, [2]Catenane-based solid state electronically reconfigurable switch. *Science*, **289**, 1172–1175 (2000).

22. W. B. Davis, W. A. Svec, M. A. Ratner, M. R. Wasielewski, Molecular-wire behaviour in p-phenylenevinylene oligomers. *Nature*, **396**, 60–63 (1998).

23. D. H. Waldeck, D. N. Beratan, Molecular electronics: Observation of molecular rectification. *Science*, **261**, 576–577 (1993).

24. R. M. Metzger, Electrical rectification by a molecule: The advent of unimolecular electronic devices. *Acc. Chem. Res.*, **32**, 950–957 (1999).

25. J. Chen, M. A. Reed, A. M. Rawlett, J. M. Tour, Large on-off ratios and negative differential resistance in a molecular electronic device. *Science*, **286**, 1550–1552 (1999).

26. S. Prathapan, S. I. Yang, J. Seth, M. A. Miller, D. F. Bocian, D. Holten. J. S. Lindsey, Synthesis and excited-state photodynamics of perylene-porphyrin dyads. 1: Parallel energy and charge transfer via a diphenylethyne linker. *J. Phys. Chem. B*, **105**, 8237–8248 (2001).

27. S. I. Yang, R. K. Lammi, S. Prapanthan, M. A. Miller, J. Seth, J. R. Diers, D. F. Bocian, J. S. Lindser, D. Holten, Synthesis and excited-state photodynamics of perylene-porphyrin dyads. Part 3. Effects of perylene, linker, and connectivity on ultrafast energy transfer. *J. Mater. Chem.*, **11**, 2420–2430 (2001).

28. M. del R. Benites, T. E. Johnson, S. Weghorn, L. Yu, P.D. Rao, J. R. Diers, S. I. Yang, C. Kirmaier, D. F. Bocian, D. Holten, J, S. Lyndsey, Synthesis and properties of weakly

coupled dendrimeric multiporphyrin light-harvesting arrays and hole-storage reservoirs. *J. Mater. Chem.*, **12**, 65–80 (2002).

29. D. Gosztola, M. P. Niemczyk, M. R. Wasielewski, Picosecond molecular switch based on bidirectional inhibition of photoinduced electron transfer using photogenerated electric fields. *J. Am. Chem. Soc.*, **120**, 5118–5119 (1998).

30. A. S. Lukas, P. J. Bushard, M. R. Wasielewski, Ultrafast molecular logic gate based on optical switching between two long-lived radical ion pair states. *J. Am. Chem. Soc.*, **123**, 2440–2441 (2001).

31. M. Kotera, J.-M. Lehn, J.-P. Vigneron, Design and synthesis of complementary components for the formation of self-assembled supramolecular rigid rods. *Tetrahedron*, **51**, 1953–1972 (1995).

32. J. L. Sessler, B. Wang, A. Harriman, Photoinduced energy transfer in associated, but noncovalently-linked photosynthetic model systems. *J. Am. Chem. Soc.*, **117**, 704–714 (1995).

33. S. L. Springs, A. Andrievsky, V. Král, J. L. Sessler, Energy transfer in a supramolecular complex assembled via sapphyrin dimer-mediated dicarboxylate anion chelation. *J. Porphyrins Phthalocyanines*, **2**, 315–325 (1998).

34. T. Arimura, S. Ide, H. Sugihara, S. Murata, J. L. Sessler, A non-covalent assembly for electron transfer based on a calixarene-porphyrin conjugate: Tweezers for a quinone. *New J. Chem.*, **23**, 977–979 (1999).

35. M. Shimomura, O. Karthaus, K. Ijiro, Tailoring of stacked π-electron arrays from electron — and/or energy donor — acceptor molecules based on two-dimensional supramolecular assemblies. *Synth. Met.*, **81**, 251–257 (1996).

36. R. A. P. Zangmeister, P. E. Smolenyak, A. S. Drager, D. F. O'Brie, N. R. Armstrong, Transfer of rodlike aggregate phthalocyanines to hydrophobized gold and silicon surfaces: Effect of phenyl-terminated surface modifiers on thin film transfer efficiency and molecular orientation. *Langmuir*, **17**, 7071–7078 (2001).

37. C. A. Mirkin, M. A. Ratner, Molecular electronics. *Annu. Rev. Phys. Chem.*, **43**, 719–754 (1992).

38. G. P. Wiederrecht, B. A. Yoon, W. A. Svec, M. R. Wasielewski, Photorefractivity in nematic liquid crystals containing electron donor-acceptor molecules that undergo intramolecular charge separation. *J. Am. Chem. Soc.*, **119**, 3358–3364 (1997).

39. A. M. Ramos, M. T. Rispens, J. K. K. Van Duren, J. C. Hummelen, R. A. J. Janssen, Photoinduced electron transfer and photovoltaic devices of a conjugated polymer with pendant fullerenes [4]. *J. Am. Chem. Soc.*, **123**, 6714–6715 (2001).

40. B. A. Diner, G. T. Babcock, G. T, Structure, Dynamics, and energy conversion efficiency in photosystem II, Edited by D. R. Ort, C. F. Yocum, *Oxygenic Photosynthesis: The Light Reactions*, Springer Netherlands, (1996). p. 213–247.

41. M. Kasha, H. R. Rawls, M. Ashraf El-Bayoumi, The exciton model in molecular spectroscopy. *Pure Appl. Chem.*, **11**, 371–392 (1965).

42. S. Das, D. P. Chatterjee, R. Ghosh, A. K. Nandi, Water soluble polythiophenes: Preparation and applications. *RSC Adv.*, **5**, 20160–20177 (2015).

43. H.-A. Ho, M. Boissinot, M. G. Bergeron, G. Corbeil, K. Doré, D. Boudreau, M. Leclerc, Colorimetric and fluorometric detection of nucleic acids using cationic polythiophene derivatives. *Angew. Chemie Int. Ed.*, **41**, 1548–1551 (2002).

44. O. Inganäs, W.R. Salaneck, J.-E. Österholm, J. Laakso, Thermochromic and solvatochromic effects in poly(3-hexylthiophene). *Synth. Met.*, **22**, 395–406 (1988).

45. J. Wang, Q. Zhang, K. J. Tan, Y. F. Long, J. Ling, C. Z. Huang, Observable temperature-dependent compaction-decompaction of cationic polythiophene in the presence of iodide. *J. Phys. Chem. B*, **115**, 1693–1697 (2011).

46. X. Liu, Y. Tang, L. Wang, J. Zhang, S. Song, C. Fan, S. Wang, Optical detection of mercury(II) in aqueous solutions by using conjugated polymers and label-free oligonucleotides. *Adv. Mater.*, **19**, 1471–1474 (2007).

47. M. Knaapila, T. Costa, V. M. Garamus, M. Kraft, M. Drechsler, U. Scherf, H. D. Burrows, Conjugated polyelectrolyte (CPE) Poly{3-[6-(N-methylimidazolium)hexyl]-2, 5-thiophene} complexed with DNA: Relation between colloidal level solution structure and chromic effects. *Macromolecules*, **47**, 4017–4027 (2014).

48. M. Knaapila, T. Costa, V. M. Garamus, M. Kraft, M. Drechsler, U. Scherf, H. D. Burrows, Polyelectrolyte complexes of a cationic all conjugated fluorene–thiophene diblock copolymer with aqueous DNA. *J. Phys. Chem. B*, **119**, 3231–3241 (2015).

49. M. Knaapila, R. C. Evans, V. M. Garamus, L. Almásy, N. K. Székely, A. Gutacker, U. Scherf, Structure and 'surfactochromic' properties of conjugated polyelectrolyte (CPE): Surfactant complexes between a cationic polythiophene and SDS in water. *Langmuir*, **26**, 15634–15643 (2010).

50. M. Knaapila, R. C. Evans, A. Gutacker, V. M. Garamus, N. K. Székely, U. Scherf, H. D. Burrows, Conjugated polyelectrolyte (CPE) poly[3-[6-(N-methylimidazolium)hexyl]-2,5-thiophene] complexed with aqueous sodium dodecylsulfate amphiphile: Synthesis, solution structure and 'surfactochromic' properties. *Soft Matter*, **7**, 6863–6872 (2011).

51. J. J. Lavigne, D. L. Broughton, J. N. Wilson, B. Erdogan, U. H. F. Bunz, 'Surfactochromic' conjugated polymers: Surfactant effects on sugar-substituted PPEs. *Macromolecules*, **36**, 7409–7412 (2003).

52. T. Costa, D. de Azevedo, B. Stewart, M. Knaapila, A. J. M. Valente, M. Kraft, U. Scherf, H. D. Burrows, Interactions of a zwitterionic thiophene-based conjugated polymer with surfactants. *Polym. Chem.*, **6**, 8036–8046 (2015).

53. K. P. R. Nilsson, J. Rydberg, L. Baltzer, O. Inganäs, Self-assembly of synthetic peptides control conformation and optical properties of a zwitterionic polythiophene derivative. *Proc. Natl. Acad. Sci. USA.* **100**, 10170–4 (2003).

54. K. F. Karlsson, P. Åsberg, K. P. R. Nilsson, O. Inganäs, Interactions between a zwitterionic polythiophene derivative and oligonucleotides as resolved by fluorescence resonance energy transfer. *Chem. Mater.*, **17**, 4204–4211 (2005).

55. E. Elmalem, F. Biedermann, M. R. J. Scherer, A. Koutsioubas, C. Toprakcioglu, G. Biffi, A. T. S. Huck, Mechanically strong, fluorescent hydrogels from zwitterionic, fully π-conjugated polymers. *Chem. Commun.*, **50**, 8930–8933 (2014).

56. S. Wang, G. C. Bazan, Solvent-dependent aggregation of a water-soluble poly(fluorene) controls energy transfer to chromophore-labeled DNA. *Chem. Commun.*, **21**, 2508–2509 (2004).

57. A. Gutacker, S. Adamczyk, A. Helfer, L. E. Garner, R. C. Evans, S. M. Fonseca, M. Knaapila, G. C. Bazan, H. D. Burrows, U. Scherf, All-conjugated polyelectrolyte block copolymers. *J. Mater. Chem.*, **20**, 1423–1430 (2010).

58. A. Gutacker, N. Koenen, U. Scherf, S. Adamczyk, J. Pina, S. M. Fonseca, A. J. M. Valente, R. C. Evans, J. Seixas de Melo, H. D. Burrows, M. Knaapila, Cationic fluorene-thiophene

diblock copolymers: Aggregation behaviour in methanol/water and its relation to thin film structures. *Polymer*, **51**, 1898–1903 (2010).

59. M. Knaapila, R. C. Evans, A. Gutacker, V. M. Garamus, M. Torkkeli, S. Adamczyk, M. Forster, U. Scherf, H. D. Burrows, Solvent dependent assembly of a polyfluorene-polythiophene 'rod-rod' block copolyelectrolyte: Influence on photophysical properties. *Langmuir*, **26**, 5056–5066 (2010).

60. S.-C. Liao, C.-S. Lai, D.-D. Yeh, M. Habibur Rahman, C.-S. Hsu, H.-L. Chen, S.-A. Chen, Supramolecular structures of an amphiphilic hairy-rod conjugated copolymer bearing poly(ethylene oxide) side chain. *React. Funct. Polym.,* **69**, 498–506 (2009).

61. B. Stewart, H. D. Burrows, Molecular dynamics study of self-assembly of aqueous solutions of poly[9,9-bis(4-sulfonylbutoxyphenylphenyl) fluorene-2,7-diyl-2,2'-bithiophene] (PBS-PF2T) in the presence of pentaethylene glycol monododecyl ether ($C_{12}E_5$). *Materials*, **9**, 379 (2016).

62. B. Liu, G. C. Bazan, *Conjugated Polyelectrolytes: Fundamentals and Applications.* Wiley-VCH Verlag GmbH & Co. KGaA, (2012).

63. H. Jiang, P. Taranekar, J. R. Reynolds, K. S. Schanze, Conjugated polyelectrolytes: Synthesis, photophysics, and applications. *Angew. Chem. Int. Ed.*, **48**, 4300–4316 (2009).

64. G. D. Joly, L. Geiger, S. E. Kooi, T. M Swager, Highly effective water-soluble fluorescence quenchers of conjugated polymer thin films in aqueous environments. *Macromolecules*, **39**, 7175–7177 (2006).

65. B. L. Nguyen, J. E. Jeong, I. H. Jung, B. Kim, V. S. Le, I. Kim, K. Khym, H. Y. Woo, Conjugated polyelectrolyte and aptamer based potassium assay via single- and two-step fluorescence energy transfer with a tunable dynamic detection range. *Adv. Funct. Mater.*, **24**, 1748–1757 (2014).

66. T. Wågberg, B. Liu, G. Orädd, B. Eliasson, L. Edman, Cationic polyfluorene: Conformation and aggregation in a 'good' solvent. *Eur. Polym. J.*, **45**, 3230–3235 (2009).

67. H. Wang, P. Lu, B. Wang, S. Qiu, M. Liu, M. Hanif, G. Cheng, S. Liu, Y. Ma, A water-soluble? π-conjugated polymer with up to 100 mg mL^{-1} solubility. *Macromol. Rapid Commun.*, **28**, 1645–1650 (2007).

68. C. Hoven, R. Yang, A. Garcia, A. J. Heeger, T.-Q. Nguyen, G. C. Bazan, Ion motion in conjugated polyelectrolyte electron transporting layers. *J. Am. Chem. Soc.*, **129**, 10976–10977 (2007).

69. J. H. Seo, A. Gutacker, Y. Sun, H. Wu, F. Huang, Y. Cao, U. Scherf, A. J. Heeger, G. C. Bazan, Improved high-efficiency organic solar cells via incorporation of a conjugated polyelectrolyte interlayer. *J. Am. Chem. Soc.*, **133**, 8416–8419 (2011).

70. D. T. McQuade, A. E. Pullen, T. M. Swager, Conjugated polymer-based chemical sensors. *Chem. Rev.*, **100**, 2537–2574 (2000).

71. S. W. Thomas, G. D. Joly, T. M. Swager, Chemical sensors based on amplifying fluorescent conjugated polymers. *Chem. Rev.*, **107**, 1339–1386 (2007).

72. M. Surin, P. G. A. Janssen, R. Lazzaroni, P. Leclére, E. W. Meijer, A. P. H. J. Schenning, Supramolecular organization of ssdna-templated p-conjugated oligomers via hydrogen bonding. *Adv. Mater.*, **21**, 1126–1130 (2009).

73. J. H. Wosnick, C. M. Mello, T. M. Swager, Synthesis and application of poly(phenylene ethynylene)s for bioconjugation: A conjugated polymer-based fluorogenic probe for proteases. *J. Am. Chem. Soc.*, **127**, 3400–3405 (2005).

74. H. D. Burrows, V. M. M. Lobo, J. Pina, M. L. Ramos, J. Seixas de Melo, A. J. M. Valente, M. J. Tapia, S. Pradhan, U. Scherf, Fluorescence enhancement of the water-soluble poly{1,4-phenylene-9,9-bis(4-phenoxybutylsulfonate)fluorene-2,7-diyl} copolymer in n-dodecylpentaoxyethylene glycol ether micelles. *Macromolecules.*, **37**, 7425–7427 (2004).

75. B. Kronberg, B. Lindman, *Surfactants and Polymers in Aqueous Solution.* John Wiley & Sons Ltd., Chichester. (2003).

76. M. Knaapila, S. M. Fonseca, B. Stewart, M. Torkkeli, J. Perlich, S. Pradhan, U. Scherf, R. A. E. Castro, H. D. Burrows, Nanostructuring of the conjugated polyelectrolyte poly[9,9-bis(4-sulfonylbutoxyphenyl) fluorene-2,7-diyl-2,2'-bithiophene] in liquid crystalline $C_{12}E_4$ in bulk water and aligned thin films. *Soft Matter*, **10**, 3103–3111 (2014).

77. M. L. Ramos, M. M. Caldeira, V. M. S. Gil, NMR study of the complexation of D-galactonic acid with tungsten (VI) and molybdenum (VI). *Carbohydr. Res.*, **297**, 191–200 (1997).

78. L. L. G. Justino, M. L. Ramos, M. M. Caldeira, V. M. S. Gil, Peroxovanadium(V) complexes of L-lactic acid as studied by NMR spectroscopy. *Eur. J. Inorg. Chem.*, **2000**, 1617–1621 (2000).

79. M. M. Caldeira, M. L. Ramos, A. M. Cavaleiro, V. M. S. Gil, Multinuclear NMR study of vanadium(V) complexation with tartaric and citric acids. *J. Mol. Struct.*, **174**, 461–466 (1988).

80. M. L. Ramos, M. M. Pereira, A. M. Beja, M. R. Silva, J. A. Paixao, V. M. S. Gil, NMR and X-ray diffraction studies of the complexation of D-(-)quinic acid with tungsten(vi) and molybdenum(vi). *J. Chem. Soc. Dalt. Trans.*, **10**, 2126–2131 (2002).

81. M. L. Ramos, L. L. G. Justino, H. D. Burrows, Structural considerations and reactivity of peroxocomplexes of V(V), Mo(VI) and W(VI). *Dalt. Trans.*, **40**, 4374–4383 (2011).

82. C. W. Tang, S. A. Vanslyke, Organic electroluminescent diodes. *Appl. Phys. Lett.*, **51**, 913–915 (1987).

83. M. L. Ramos, L. L. G. Justino, A. I. N. Salvador, A. R. E. de Sousa, P. E. Abreu, S. M. Fonseca, H. D. Burrows, NMR, DFT and luminescence studies of the complexation of Al(III) with 8-hydroxyquinoline-5-sulfonate. *Dalt. Trans.*, **41**, 12478–12489 (2012).

84. M. L. Ramos, A. R. E. de Sousa, L. L. G. Justino, S. M. Fonseca, C. F. G. C. Geraldes, H. D. Burrows, Structural and photophysical studies on gallium(III) 8-hydroxyquinoline-5-sulfonates. Does excited state decay involve ligand photolabilisation? *Dalt. Trans.*, **42**, 3682–3694 (2013).

85. M. L. Ramos, L. L. G. Justino, A. Branco, C. M. G. Duarte, P. E. Abreu. S. M. Fonseca, H. D. Burrows, NMR, DFT and luminescence studies of the complexation of Zn(II) with 8-hydroxyquinoline-5-sulfonate. *Dalt. Trans.*, **40**, 11732–11741 (2011).

86. M. L. Ramos, L. L. G. Justino, A, Branco, S.M. Fonseca, H. D. Burrows, Theoretical and experimental insights into the complexation of 8-hydroxyquinoline-5-sulfonate with divalent ions of Group 12 metals. *Polyhedron*, **52**, 743–749 (2013).

87. M. L. Ramos, L. L. G. Justino, S. M. Fonseca, H. D. Burrows, NMR, DFT and luminescence studies of the complexation of V(V) oxoions in solution with 8-hydroxyquinoline-5-sulfonate. *New J. Chem.*, **39**, 1488–1497 (2015).

88. M. L. Ramos, L. L. G. Justino, P. E. Abreu, S. M. Fonseca, H. D. Burrows, Oxocomplexes of Mo(VI) and W(VI) with 8-hydroxyquinoline-5-sulfonate in solution: Structural studies

and the effect of the metal ion on the photophysical behaviour. *Dalt. Trans.*, **44**, 19076–19089 (2015).

89. E. Bardez, I. Devol, B. Larrey, B. Valeur, Excited-state processes in 8-hydroxyquinoline: Photoinduced tautomerization and solvation effects. *J. Phys. Chem. B*, **101**, 7786–7793 (1997).

90. A. Curioni, M. Boero,W.Andreoni, Alq3: Ab initio calculations of its structural and electronic properties in neutral and charged states. *Chem. Phys. Lett.*, **294**, 263–271 (1998).

91. C. H. Chen, J. Shi, Metal chelates as emitting materials for organic electroluminescence. *Coord. Chem. Rev.*, **171**, 161–174 (1998).

92. R. Ballardini, G. Varani, M. T. Indelli, F. Scandola, phosphorescent 8-quinolinol metal chelates. Excited-state properties and redox behavior. *Inorg. Chem.*, **25**, 3858–3865 (1986).

93. Z. Zhujun, W. R. Seitz, A fluorescent sensor for aluminum(III), magnesium(II), zinc(II) and cadmium(II) based on electrostatically immobilized quinolin-8-ol sulfonate. *Anal. Chim. Acta*, **171**, 251–258 (1985).

94. H. D. Burrows, T. Costa, M. L. Ramos, A. J. M. Valente, B. Stewart, L. L. G. Justino, A. I. A. Almeida, N. L. Catarina, R. Mallavia, M. Knaapila, Self-assembled systems of water soluble metal 8-hydroxyquinolates with surfactants and conjugated polyelectrolytes. *Phys. Chem. Chem. Phys.*, **18**, 16629–16640 (2016).

95. H. D. Burrows, M. Knaapila, A. P. Monkman, M. J. Tapia, S. M. Fonseca, M. L. Ramos, W. Pyckhout-Hintzen, S. Pradhan, U. Scherf, Structural studies on cationic poly{9,9-bis[6-(N,N,N-trimethylammonium)alkyl]fluorene-co-1,4-phenylene} iodides in aqueous solutions in the presence of the non-ionic surfactant pentaethyleneglycol monododecyl ether ($C_{12}E_5$). *J. Phys. Condens. Matter*, **20**, 104210 (2008).

96. H. Christenson, S. E. Friberg, Spectroscopic investigation of the mutual interactions between nonionic surfactant, hydrocarbon, and water. *J. Colloid Interface Sci.*, **75**, 276–285 (1980).

97. Z. Wu, J. C. Grossman, Prediction of ultra-high aspect ratio nanowires from self-assembly. *Nano Lett.*, **8**, 2697–2705 (2008).

98. A. P. Alivisatos, Semiconductor clusters, nanocrystals, and quantum dots. *Science*, **271**, 933–937 (1996).

99. X. Duan, Y. Huang, R. Agarwal, C. M. Lieber, Single-nanowire electrically driven lasers. *Nature*, **421**, 241–245 (2003).

100. Y. Cui, Q. Wei, H. Park, C. M. Lieber, Nanowire nanosensors for highly sensitive and selective detection of biological and chemical species. *Science*, **293**, 1289–1292 (2001).

101. W. U. Huynh, J. J. Dittmer, A. P. Alivisatos, Hybrid nanorod-polymer solar cells. *Science*, **295**, 2425–2427 (2002).

102. M. Artemyev, B. Möller, U. Woggon, Unidirectional alignment of CdSe nanorods. *Nano Lett.*, **3**, 509–512 (2003).

103. O. Harnack, C. Pacholski, H. Weller, A. Yasuda, J. M. Wessels, Rectifying behavior of electrically aligned ZnO nanorods. *Nano Lett.*, **3**, 1097–1101 (2003).

104. S.-M. Koo, Q. Li, M. D. Edelstein, C. A. Richter, E. M. Vogel, Enhanced channel modulation in dual-gated silicon nanowire transistors. *Nano Lett.*, **5**, 2519–2523 (2005).

105. D. Wang, F. Qian, C. Yang, Z. Zhong, C. M. Lieber, Rational growth of branched and hyperbranched nanowire structures. *Nano Lett.*, **4**, 871–874 (2004).

106. Y. H. Tang, T. K. Sham, A. Jürgensen, Y. F. Hu, C. S. Lee, S. T. Lee, Phosphorus-doped silicon nanowires studied by near edge x-ray absorption fine structure spectroscopy. *Appl. Phys. Lett.*, **80**, 3709 (2002).

107. J. Westwater, Growth of silicon nanowires via gold/silane vapor–liquid–solid reaction. *J. Vac. Sci. Technol. B Microelectron. Nanom. Struct.*, **15**, 554 (1997).

108. R. Juhasz, N. Elfström, J. Linnros, Controlled fabrication of silicon nanowires by electron beam lithography and electrochemical size reduction. *Nano Lett.*, **5**, 275–280 (2005).

109. E. J. Menke, M. A. Thompson, C. Xiang, L. C. Yang, R. M. Penner, Lithographically patterned nanowire electrodeposition. *Nat. Mater.*, **5**, 914–919 (2006).

110. A. Morales, C. M. Lieber, A laser ablation method for the synthesis of crystalline semiconductor nanowires. *Science*, **279**, 208–11 (1998).

111. Y. Ding, Y. Mei, J. Z. H. Zhang, F. M. Tao, Efficient bond function basis set for π-π interaction energies. *J. Comput. Chem.*, **29**, 275–279 (2008).

112. Y. F. Zhang, Y. H. Tang, N. Wang, D. P. Yu, C. S. Lee, I. Bello, S. T. Lee, Silicon nanowires prepared by laser ablation at high temperature. *Appl. Phys. Lett.*, **72**, 1835–1837 (1998).

113. R. S. Wagner, W. C. Ellis, Vapor-liquid-solid mechanism of single crystal growth. *Appl. Phys. Lett.*, **4**, 89–90 (1964).

114. E. I. Givargizov, Fundamental aspects of VLS growth. *J. Cryst. Growth*, **31**, 20–30 (1975).

115. R. D. Kamien, Topology from the bottom up. *Science*, **299**, 1671–1673 (2003).

116. C. B. Murray, C. R. Kagan, M. G. Bawendi, Syntheses and characterization of monodisperse nanocrytals and close-packed nanocrystal assemblies. *Annu. Rev. Mater. Sci.*, **30**, 545–610 (2000).

117. B. D. Busbee, S. O. Obare, C. J. Murphy, An improved synthesis of high-aspect-ratio gold nanorods. *Adv. Mater.*, **15**, 414–416 (2003).

118. Y. Sun, Y. Xia, Shape-controlled synthesis of gold and silver nanoparticles. *Science*, **298**, 2176-2179 (2002).

119. F. M. van der Kooij, K. Kassapidou, H. N. W. Lekkerkerker, H. N. W. Liquid crystal phase transitions in suspensions of polydisperse plate-like particles. *Nature*, **406**, 868–871 (2000).

120. J. C. Love, A. R. Urbach, M. G. Prentiss, G. M. Whitesides, Three-dimensional self-assembly of metallic rods with submicron diameters using magnetic interactions. *J. Am. Chem. Soc.*, **125**, 12696–12697 (2003).

121. Z. Tang, N. A. Kotov, M. Giersig, Spontaneous organization of single CdTe nanoparticles into luminescent nanowires. *Science*, **297**, 237–240 (2002).

122. Z. Tang, Z, Zhang, Y. Wang, S. C. Glotzer, N. A. Kotov, Self-assembly of CdTe nanocrystals into free-floating sheets. *Science*, **314**, 274–278 (2006).

123. Z. R. Tian, J. Liu, H. Xu, J. A. Voigt, B. Mckenzie, C. M. Matzke, Shape-selective growth, patterning, and alignment of cubic nanostructured crystals via self-assembly. *Nano Lett.*, **3**, 179–182 (2003).

124. H. F. Zhang, A. C. Dohnalkova, C. M. Wang, J. S. Young, E. C. Buck, L. S. Wang, Lithium-assisted self-assembly of aluminum carbide nanowires and nanoribbons. *Nano Lett.*, **2**, 105–108 (2002).

125. M. Chen, P. C. Searson, The dynamics of nanowire self-assembly. *Adv. Mater.*, **17**, 2765–2768 (2005).

126. T. D. Clark, J. Tien, D. C. Duffy, K. E. Paul, G. M. Whitesides, Self-assembly of 10-μm-sized objects into ordered three-dimensional arrays. *J. Am. Chem. Soc.*, **123**, 7677–7682 (2001).

127. R. Knischka, F. Dietsche, R. Hanselmann, H. Frey, R. Mülhaupt, P. J. Lutz, Silsesquioxane-based amphiphiles. *Langmuir*, **15**, 4752–4756 (1999).

128. C. M. Leu, G. M. Reddy, K. H. Wei, C. F. Shu, Synthesis and dielectric properties of polyimide-chain-end tethered polyhedral oligomeric silsesquioxane nanocomposites. *Chem. Mater.*, **15**, 2261–2265 (2003).

129. J. L. Gray, S. Atha, R. Hull, J. A. Floro, Hierarchical self-assembly of epitaxial semiconductor nanostructures. *Nano Lett.*, **4**, 2447–2450 (2004).

130. S. Pierrat, I. Zins, A. Breivogel, G. Sönnichsen, G. Self-assembly of small gold colloids with functionalized gold nanorods. *Nano Lett.*, **7**, 259–263 (2007).

131. Z. Zhang, M. A. Horsch, M. H. Lamm, S. C. Glotzer, Tethered nano building blocks: Toward a conceptual framework for nanoparticle self-assembly. *Nano Lett.*, **3**, 1341–1346 (2003).

132. M. A. Horsch, Z. Zhang, S. C Glotzer, Self-assembly of laterally-tethered nanorods. *Nano Lett.*, **6**, 2406–2413 (2006).

133. Z. Zhang, S. C. Glotzer, Self-assembly of patchy particles. *Nano Lett.*, **4**, 1407–1413 (2004).

134. Z. Zhang, Z. Tang, N. A. Kotov, S. C. Glotzer, Simulations and analysis of self-assembly of CdTe nanoparticles into wires and sheets. *Nano Lett.*, **7**, 1670–1675 (2007).

135. M. H. Lamm, T. Chen, S. C. Glotzer, Simulated assembly of nanostructured organic/inorganic networks. *Nano Lett.*, **3**, 989–994 (2003).

136. H. Li, W. Zhang, W. Xu, X. Zhang, Hydrogen bonding governs the elastic properties of poly(vinyl alcohol) in water: Single-molecule force spectroscopic studies of PVA by AFM. *Macromolecules*, **33**, 465–469 (2000).

137. X. Zhang, E. R. Chan, S. C. Glotzer, Self-assembled morphologies of monotethered polyhedral oligomeric silsesquioxane nanocubes from computer simulation. *J. Chem. Phys.*, **123**, 184718 (2005).

138. M. A. Horsch, Z. Zhang, S. C. Glotzer, Self-assembly of polymer-tethered nanorods. *Phys. Rev. Lett.*, **95**, 1–4 (2005).

139. E. Jiménez-Millán, J. J. Giner-Casares, E. Muñoz, M. T. Martín-Romero, L. Camacho, Self-assembly of acridine orange into H-aggregates at the air/water interface: Tuning of orientation of headgroup. *Langmuir*, **27**, 14888–14899 (2011).

140. G. B. Rocha, R. O. Freire, A. M. Simas, J. J. P. Stewart, RM1: A reparameterization of AM1 for H, C, N, O, P, S, F, Cl, Br, and I. *J. Comput. Chem.*, **27**, 1101–1111 (2006).

141. G. O. Fruhwirth, L. P. Fernandes, G. Weitsman, G. Patel, M. Kelleher, K. Lawler, A. Brock, S. P. Poland, D. R. Matthews, G. Kéri, P. R. Barber, B. Vojnovic, S. M. Ameer-Beg, A. C. C. Coolen, F. Fraternali, T. Ng, How Förster resonance energy transfer imaging improves the understanding of protein interaction networks in cancer biology. *Chem. Phys. Chem.*, **12**, 442–461 (2011).

142. F. T. S. Chan, C. F. Kaminski, G. S. Kaminski Schierle, HomoFRET fluorescence anisotropy imaging as a tool to study molecular self-assembly in live cells. *Chem. Phys. Chem.*, **12**, 500–509 (2011).

143. R. Rieger, A. Kobitski, H. Sielaff, G. U. Nienhaus, Evidence of a folding intermediate in RNase H from single-molecule FRET experiments. *Chem. Phys. Chem.*, **12**, 627–633 (2011).

144. L. Chen, Y. Honsho, S. Seki, D. Jiang, Light-harvesting conjugated microporous polymers: Rapid and highly efficient flow of light energy with a porous polyphenylene framework as antenna. *J. Am. Chem. Soc.*, **132**, 6742–6748 (2010).

145. J.-F. Yin, J.-G, Chen, Z.-Z. Lu, K.-C. Ho, H.-C. Lin, K.-L. Lu, Toward optimization of oligothiophene antennas: New ruthenium sensitizers with excellent performance for dye-sensitized solar cells. *Chem. Mater.*, **22**, 4392–4399 (2010).

146. N. Robertson, Optimizing dyes for dye-sensitized solar cells. *Angew. Chemie Int. Ed.*, **45**, 2338–2345 (2006).

147. D. P. Hagberg, T. Edvinsson, T. Marinado, G. Boschloo, A. Hagfeldt, L. Sun, A novel organic chromophore for dye-sensitized nanostructured solar cells. *Chem. Commun.*, **21**, 2245–2247 (2006).

148. W. Ma, C. Yang, X. Gong, K. Lee, A. J. Heeger, Thermally Stable, Efficient polymer solar cells with nanoscale control of the interpenetrating network morphology. *Adv. Funct. Mater.*, **15**, 1617–1622 (2005).

149. S. Kim, J. K. Lee, S. O. Kang, J. Ko, J.-H. Yum, S. Fantacci, F. De Angelis, D. Di Censo, M. K. Nazeeruddin, M. Grätzel, Molecular engineering of organic sensitizers for solar cell applications. *J. Am. Chem. Soc.*, **128**, 16701–16707 (2006).

150. O. Valdes-Aguilera, D. C. Neckers, Aggregation phenomena in xanthene dyes. *Acc. Chem. Res.*, **22**, 171–177 (1989).

151. S. Kim, T. K. An, J. Chen, I. Kang, S. H. Kang, D. S. Chung, C. E. Park, Y.-H. Kim, S.-K Kwon, H-aggregation strategy in the design of molecular semiconductors for highly reliable organic thin film transistors. *Adv. Funct. Mater.*, **21**, 1616–1623 (2011).

152. A. Ulman, *An Introduction to Ultrathin Organic Films: From Langmuir–Blodgett to Self-Assembly*. Academic press, (2013).

153. M. C. Petty, *Langmuir-Blodgett Films: An Introduction*. Cambridge University Press, (1996).

154. A. D. Becke, Density-functional thermochemistry. III. The role of exact exchange. *J. Chem. Phys.*, **98**, 5648 (1993).

155. C. Lee, W. Yang, R. G. Parr, Development of the Colle-Salvetti correlation-energy formula into a functional of the electron density. *Phys. Rev. B*, **37**, 785–789 (1988).

156. S. H. Vosko, L. Wilk, M. Nusair, Accurate spin-dependent electron liquid correlation energies for local spin density calculations: A critical analysis. *Can. J. Phys.*, **58**, 1200–1211 (1980).

157. P. J. Stephens, F. J. Devlin, C. F. Chabalowski, M. J. Frisch, Ab initio calculation of vibrational absorption and circular dichroism spectra using density functional force fields. *J. Phys. Chem.*, **98**, 11623–11627 (1994).

158. M. W. Schmidt, K. K. Baldridge, J. A. Boatz, S. T. Elbert, M. S. Gordon, J. H. Jensen, S. Koseki, N. Matsanuga, K. A. Nguyen, S. Su, T. L. Windus, M. Dupuis, J. A. Montgomery, General atomic and molecular electronic structure system. *J. Comput. Chem.*, **14**, 1347–1363 (1993).

159. C. J. Cramer, *Essentials of Computational Chemistry Theories and Models. Essentials of Computational Chemistry*, J. Wiley, (2002).

160. R. G. Parr, Y. Weitao, *Density-Functional Theory of Atoms and Molecules*, Oxford University Press, (1989).

161. S. Miertuš, E. Scrocco, J. Tomasi, Electrostatic interaction of a solute with a continuum. A direct utilizaion of AB initio molecular potentials for the prevision of solvent effects. *Chem. Phys.*, **55**, 117–129 (1981).

162. J. Tomasi, B. Mennucci, R. Cammi, Quantum mechanical continuum solvation models. *Chem. Rev.*, **105**, 2999–3094 (2005).

163. M. Grell, D. D. C. Bradley, X. Long, T. Chamberlain, M. Inbasekaran, E. P. Woo, M. Soliman, Chain geometry, solution aggregation and enhanced dichroism in the liquid crystalline conjugated polymer poly(9,9-dioctylfluorene). *Acta Polym.*, **49**, 439–444 (1998).

164. M. Grell, D. D. C. Bradley, G. Ungar, J. Hill, K. S. Whitehead, Interplay of physical structure and photophysics for a liquid crystalline polyfluorene. *Macromolecules*, **32**, 5810–5817 (1999).

165. J. Teetsov, M. Anne Fox, Photophysical characterization of dilute solutions and ordered thin films of alkyl-substituted polyfluorenes. *J. Mater. Chem.*, **9**, 2117–2122 (1999).

166. L. L. G. Justino, M. L. Ramos, M. Knaapila, A. T. Marques, C. J. Kudla, U. Scherf, L. Almásy, R. Schweins, H. D. Burrows, A. P. Monkman, Gel formation and interpolymer alkyl chain interactions with Poly(9,9-dioctylfluorene-2,7-diyl) (PFO) in toluene solution: Results from NMR, SANS, DFT, and semiempirical calculations and their implications for PFO β-phase formation. *Macromolecules*, **44**, 334–343 (2011).

167. J. J. P. Stewart, *MOPAC2007*. (2007).

168. T. Ikeda, T. Iijima, R. Sekiya, O. Takahashi, T. Haino, Cooperative self-assembly of carbazole derivatives driven by multiple dipole–dipole interactions. *J. Org. Chem.*, **81**, 6832–6837 (2016).

169. Y. Zhao, D. G. Truhlar, The M06 suite of density functionals for main group thermochemistry, thermochemical kinetics, noncovalent interactions, excited states, and transition elements: Two new functionals and systematic testing of four M06-class functionals and 12 other function. *Theor. Chem. Acc.*, **120**, 215–241 (2008).

170. M. J. Frisch, G. W. Trucks, H. B. Schlegel, G. E. Scuseria, M. A. Robb, J. R. Cheeseman, A. Scalmani, V. Barone, B. Mennucci, G. Petersson, *Gaussian 09W, Revision A.1*. Wallingford (CT): Gaussian. (2009).

171. H. T. Black, N, Yee, Y, Zems, D. F. Perepichka, D. F. Complementary hydrogen bonding modulates electronic properties and controls self-assembly of donor/acceptor semiconductors. *Chem. — A Eur. J.*, **22**, 17251–17261 (2016).

172. O. A. Bell, G. Wu, J, S. Haataja, F. Brömmel, N. Fey, A. M. Sneddon, R. L. Harniman, R. M. Richardson, O. Ikkala, A. Zhang, C. F. J. Faul, Self-assembly of a functional oligo(aniline)-based amphiphile into helical conductive nanowires. *J. Am. Chem. Soc.*, **137**, 14288–14294 (2015).

173. J. L. Pascual-ahuir, E. Silla, I. Tuñon, GEPOL: An improved description of molecular surfaces. III. A new algorithm for the computation of a solvent-excluding surface. *J. Comput. Chem.*, **15**, 1127–1138 (1994).

174. T. Yanai, D. P. Tew, N. C. Handy, A new hybrid exchange–correlation functional using the Coulomb-attenuating method (CAM-B3LYP). *Chem. Phys. Lett.*, **393**, 51–57 (2004).

Chapter 4

Specific Interactions Between Electroactive Conducting Polymers and DNA Bases

Carlos Alemán[*,†,‡] *and David Zanuy*[*,§]

Departament d'Enginyeria Química, ETSEIB,
Universitat Politècnica de Catalunya, Av. Diagonal 647,
E-08028 Barcelona, Spain

[†]*Center for Research in Nano-Engineering,*
Universitat Politècnica de Catalunya, Campus Sud,
Edifici C', C/Pasqual i Vila s/n,
Barcelona, E-08028, Spain
[‡]*carlos.aleman@upc.edu*
[§]*david.zanuy@upc.edu*

Understanding the molecular interactions and recognition processes between DNA and electroactive conducting polymer (ECP) chains have a high and increasing interest for potential biomedical and biotechnological applications. In fact, ECPs are among the most important organic materials due to the versatility of their response against electrochemical and electrical stimuli, environmental stability, biocompatibility and bioactivity and ease synthesis. Furthermore, ECPs can be engineered to combine their inherent electronic, electrochemical and charge transport properties and improve their ability to recognize DNA building blocks.

In this chapter, after the first section that is devoted to provide a very brief introduction to ECPs and their potential bioapplications, the second section focused on providing simple theoretical chemistry concepts that are necessary for the understanding of the next sections. Thus, subsequent sections highlight the identification of specific interactions formed between ECP and DNA molecules. Specifically, the discussion of such interactions has been organized considering separately experimental observations, theoretical predictions using sophisticated quantum mechanical calculations on small models made of building blocks and, finally, atomistic computer simulations using realistic models.

1.   Introduction

Electroactive conducting polymers (ECPs) are being widely used in biomedical science due to its inherent electronic, electrochemical and charge transport properties.[1,2] Furthermore, many ECPs are biocompatible,[3–5] which is a requisite towards the design of materials for biomedical applications. ECPs are a special class of compounds that consists of long conjugated polymeric chains in which alternation of double and single-bonded sp^2 hybridized atoms is present. This conjugation is responsible for charge mobility along the polymer backbone and between adjacent chains endowing the polymer with semiconductive and electrochemical properties.

ECPs are classified into three different groups: non-cyclic polyenes (e.g. polyacetylene), cyclic polyenes (e.g. poly(*p*-phenylene) and its derivatives), and polyheterocycles (e.g. polypyrrole (PPy), polythiophene (PTh) and polyaniline (PAni)). The last group is currently considered the most versatile due to their excellent properties, especially in terms of electrochemical and environmental stabilities. Although chemical synthesis provides many different possible routes to obtain a variety of ECPs, electrochemical synthesis is frequently the preferred alternative since this procedure is relatively straightforward and enables control over the yield, morphology and electrochemical properties of the resulting material.

Due to such outstanding properties, ECPs have been frequently used as biosensors, tissue-engineering scaffolds, neural probes, drug-delivery devices and bio-actuators, as has been recently reviewed.[6–13] Within the particular field of biosensors, the combination of experimental and theoretical approaches has shown a relationship between the capacity of ECPs to interact with corresponding analytes and their sensing capacities. This trend was clearly reflected in recent studies devoted to examining the detection of dopamine[14,15] and morphine,[16] an important neurotransmitter in the mammalian central nervous system and the principal active component of opium, respectively, using a combination of electrochemical assays and theoretical calculations on model systems. Moreover, detailed knowledge of the molecular interactions between simple ECPs and such species allowed us to design new sensors with optimized properties.[15,16] Similarly, theoretical investigations on complexes formed by model ECPs and metallic cations,[17,18] vapour of solvents,[19] and DNA bases[20,21] have been carried out to establish relationships between the experimental sensing information and both the interaction pattern and the binding energies.

The aim of this chapter is to give an overview of the molecular interactions between the most common and important ECPs and DNA, which have been investigated using different theoretical methods, and their correlation with the experimentally determined recognition processes. Understanding of such processes, especially at the molecular level, will aid in better designing new ECP-based materials for the specific detection of nucleotide sequences. The following section provides a brief description of the main theoretical methods used in such studies, which is expected to facilitate the comprehension of theoretical tools used in our different computational studies. After this, experimental results on ECP···plasmid DNA complexes are briefly presented, proving the existence of specific hydrogen bonds between such two species and giving some insights about the mechanism proceeded to form such adducts. The subsequent two sections are devoted to the explanation of the results derived from combined theoretical and spectroscopic studies on different model systems that cover small model complexes, which were constructed using the building blocks of the two interacting species, and realistic ECP···oligonucleotides complexes. The main conclusions are summarized in the last part of this chapter.

2. Theoretical Background

Given the current computing capabilities, computational chemistry can presently be considered as an important branch of chemistry. This allows the incorporation of our knowledge on theoretical chemistry into efficient computer programs, many of which are available commercially, which can be used to solve a large variety of chemical and biological problems. The range of chemical systems that can be considered is very broad, going from isolated atoms or small molecules to polymers, biological macromolecules or molecular associations in gas, liquid or solid phases, all in both static and dynamic situations.[22,23] In addition, specific graphic software is usually used to interactively visualize in an attractive way the results obtained.

Many computational studies are oriented to determine properties of systems with a fixed disposition of the constituent atoms and/or molecules. Other studies, such as conformational searches, are primarily aimed to establish which of the nuclear coordinates correspond to the stable structures of a given system. In any case, the first stage in computational studies is the selection of the theoretical methodology to be employed. Thus, chemical problems can be addressed by using quantum mechanics (QM) or by Classical Mechanics, also called molecular mechanics (MM). In the former chemical systems are described as a set of nuclei and electrons that follow quantum physics principles. In the latter, systems are considered as a set of particles whose centres of masses are located at each respective nucleus that behave as expected from classical physics theory. As a consequence of explicitly considering the electrons, QM methods are frequently used to provide a precise description of the conformational preferences of small and medium size molecules, their interaction patterns and their binding energies. Although QM studies offer the most accurate description of matter and allow evaluation of electronic-dependent properties that cannot be analyzed by other methods, their computational requirements are very high increasing rapidly with the size of the system under study. As a consequence, QM calculations are typically restricted to systems involving a 50–100 atoms, while less accurate and less computationally demanding MM methods are used for bigger systems and, in particular, for studies where the electronic structure is not generally subject to any reorganization.

QM calculations usually involve the problem of solving the time-independent Schrödinger equation associated to the system under study:

$$\hat{H}\Psi = E\Psi \tag{1}$$

where $\hat{H}$ is the Hamiltonian operator containing the kinetic and potential energy of the nuclei and electrons, E is the energy, and Ψ is the wave function that implicitly contains all the measurable information about the system. The second-order differential Schrödinger equation is an eigen-value equation, whose resolution provides pairs of $\{E, \Psi\}$ values. Between them, the pair with lower E describes the energy and the behaviour of the system in its ground state.

Solving the Schrödinger equation is in practice only affordable by using some approximations. Among them, the Born-Oppenheimer approximation and the description of ψ(electrons) by molecular orbitals are particularly outstanding. The former considers that electrons can adjust almost instantaneously to any change in the positions of the nuclei, whereas the latter expresses ψ as lineal combinations of a set of selected mathematical functions centred at the nuclei. In practice, there are two procedures, which essentially differ in their philosophies, to solve a chemical problem using QM calculations: *ab initio* and density functional theory (DFT) methodologies.

Ab initio methods are grounded on solving the Schrödinger equation without using empirical or semiempirical parameters, (i.e. from the first principles of quantum theory[24] with no inclusion of experimental data). As was stated before, some approximations are necessary to solve the mathematical problem, different *ab initio* methodologies being proposed depending on their thoroughness. Detailed description of such methods, which is out of the scope of this chapter, is provided in Ref. 24. On the other hand, DFT methods, which represent an alternative to solving the Schrödinger equation, are currently used to study the electronic structure (essentially the ground state) of a large number of molecules. Within this theory, the parameter to be optimized is the functional that relates the electronic density of the system with its energy (i.e. DFT methods work with the density functional instead of complex many-electron wavefunctions). Thus, computational requirements are small in comparison with *ab initio* methods, as the wavefunction of an N-electron system depends on $3N$ spatial coordinates, while the density has only three variables. The problem is that, although it is known that it exists, the exact form of the functional connecting the electron density and the energy of the system is unknown. The goal of the different DFT methods is to design a functional linking these two properties. Thus, the difference between various DFT methods is the choice of the functional used to represent the energy.

MM methods are based on mathematical models that assume a molecule as a collection of balls and springs representing, respectively, particles and bonds. Thus, this method ignores the electronic motions and each particle is described by its radius, hardness and net charge. The potential energy function $\Phi(\vec{R})$, or force-field, used to describe interactions among atoms in MM is based on classical mechanics and for a system of N particles has the form:

$$\Phi(\vec{R}) = \sum_{bonds} \frac{1}{2} K_b (b - b_0)^2 + \sum_{angles} \frac{1}{2} K_\theta (\theta - \theta_0)^2 + \sum_{dihedrals} K[1 + \cos(n\phi - \delta)]$$

$$+ \sum_{impropers} \frac{1}{2} K_\xi (\xi - \xi_0)^2 + \sum_{pairs(i,j)} \left(\frac{C_{12}(i,j)}{r_{ij}^{12}} - \frac{C_6(i,j)}{r_{ij}^6} + \frac{q_i q_j}{4\pi\varepsilon_0 r_{ij}} \right)$$

$$(2)$$

The first two terms represent the covalent bond stretching and bond angle energies, respectively, which are described as the elastic deformation of a spring. The following two terms describe the torsional terms: the first term, known as torsion potential, relates with the energy barriers associated with the rotation around single covalent bonds and those contributions are adjusted to a Fourier expansion of cosines. The second term takes into account the geometry distortions that rigid bond geometries can undergo, such as movements out of the plane in molecules that present conjugated π electrons. This term can be expressed in some force-fields as a Fourier expansion. The last term, the non-bonded part, models the interaction between atoms that are three or more bonds away plus all intermolecular interactions. It is composed of the van der Waals interaction, described usually through a Lennard-Jones potential, and the Coulomb interaction. The parameters K_b, b_0, K_θ, θ_0, K_ξ, ξ_0, K, n, δ, C_{12}, C_6 and q_i can be derived from QM calculations on model systems or obtained empirically.

On the other hand, MM can be used for different procedures typically associated with molecular modelling and computational chemistry, like computation of the energy of a molecular system, energy minimization, Monte Carlo and molecular dynamics (MD), the latter being the most frequent in the study of ECP⋯DNA complexes. The aim of MD is to reproduce the time-dependent motional behaviour of a molecule. It is assumed that the atoms in the molecular system interact with each other according to the expressions used in force-field (Eq. (2)). Thus, the system is described as a system of point masses moving in an effective field. In the MD method the motion of the point charges is governed by Newton's equation of motion (Eq. (3)):

$$\frac{d\vec{v}_i(t)}{dt} = m_i^{-1}\vec{F}_i(\{\vec{r}_i(t)\}) \tag{3}$$

which are integrated numerically for all particles interacting through a known potential like that displayed in Eq. (2). The particles positions and velocities are represented by $\vec{r}_i(t)$ and $\vec{v}_i(t)$, respectively. The forces, $\vec{F}_i$, are obtained from the potential functions through $\vec{F}_i = -\frac{\partial\Phi(\vec{R})}{\partial r_i}$.

3. Experimental Identification of Specific Interactions Between ECP and Plasmid DNA

Although the interaction between oxidized ECPs with DNA was traditionally attributed to the tendency of the latter biomacromolecule to interact with positively charged molecules, around one decade ago different experimental studies in ECP⋯DNA adducts were oriented toward the identification of specific interactions (i.e. those depending on the chemical environment and the spatial disposition and/or orientation of the chemical groups). Thus, the existence of specific interactions in these complexes, in particular hydrogen bonds, complementing both strong (i.e. electrostatic) and weak (i.e. stacking and van der Waals) non-specific interactions could have important implications in the use of ECPs for the detection of well-defined nucleotide sequences.

Gel electrophoresis, UV-Vis spectroscopy and circular dichroism (CD) studies were performed considering different ECP⋯DNA complexes,[25-27] which in turn were prepared by mixing different ECP: plasmid DNA mass ratios. The ECPs employed for such studies were: poly(3,4-ethylenedioxythiophene)[25] (PEDOT), poly(3-methylthiophene)[25] (P3TM), poly(3-methylthiophene-*co*-3,4-ethylendioxythiophene)[25]   (P(3MT-*co*-ED OT)), poly(3-thiophen-3-yl-acrylic acid methyl ester)[26] (PT3AME), poly(2-thiophen-3-yl-malonic acid dimethyl ester)[26] (PT3MDE), PPy,[27] poly(N-

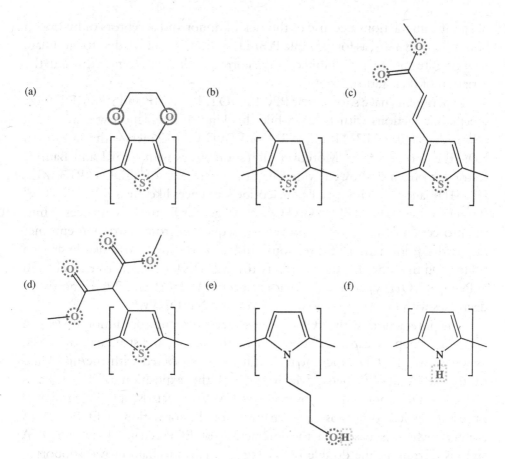

Figure 1: Chemical structure of the ECPs used to study the possible formation of ECP-DNA adducts: (a) poly(3,4-ethylenedioxythiophene) (PEDOT), (b) poly(3-methylthiophene) (P3TM), (c) poly(3-thiophen-3-yl-acrylic acid methyl ester) (PT3AME), (d) poly(2-thiophen-3-yl-malonic acid dimethyl ester) (PT3MDE), (e) polypyrrole (PPy) and (f) poly(N-(hydroxypropyl)pyrrole) (PNPrOHPy). Functional groups able to act as donors of hydrogen bonds are encircled with plain lines and those able to act as acceptors are remarked by dashed squares. Sulphur groups are encircled by dashed lines since depending on the oxidation state of the polymer can become acceptors, as well.

(hydroxypropyl)pyrrole)[27] (PNPrOHPy) and poly(pyrrole-*co*-N-(hydroxypropyl)pyrrole))[27] (P(Py-*co*-NPrOHPy)).

Inspection of the chemical structure of such ECPs (Fig. 1) shows very different chemical characteristics depending on the contained functional groups. For example, the oxygen atoms of the dioxane ring in PEDOT and the ester side groups in both PT3AME and PT3MDE act as excellent acceptors of hydrogen bonds, while N–H of PPy tend to act as a hydrogen bond donor. In opposition, the chemical structure of P3TM is not suitable for the formation of specific interactions because of the lack of donor and acceptors of hydrogen bonds. Copolymers, as for example P(3MT-*co*-EDOT), allowed us to modulate the opposite tendencies exhibited by the corresponding monomers through the variation of their ratios.

Experiments have shown that PPy, PEDOT, PT3AME and PT3MDE form specific interactions with DNA, while this kind of interaction was weak and poorly detected for P3MT and P(3M-*co*-EDOT) copolymers with low ratio of EDOT monomers.[25–27] Moreover, digestion assays with EcoRI and BamHI restriction enzymes showed evidence of the interaction of PEDOT, PT3AME, PT3MDE and PPy with specific nucleotide sequences like the 5′-G/AATTC-3′ (target for EcoRI) and 5′-G/GATCC-3′ (target for BamHI) sequences. Thus, the protection of these target nucleotide sequences from restriction enzymes required the formation of directional and specific interactions like hydrogen bonds. On the other hand, the affinity towards DNA of P(3M-*co*-EDOT) and P(Py-*co*-NPrOHPy) copolymers with respect to PEDOT and PPy, respectively, decreases with increasing amount of 3MT and NPrOHPy.[25,27]

The interaction with ECPs bearing acceptor and/or donor hydrogen bonding groups creates an alteration in the secondary structure of DNA, as was shown by CD experiments.[26] This was associated with the unfolding of the DNA double helix, which explains the exposition of DNA bases in ECP···DNA adducts as observed by UV-Vis spectroscopy. Experimental observations led to propose a mechanism for the formation of ECP···DNA adducts, which is based on the interaction of ECPs chains between DNA strands disrupting the double helix (Fig. 2). This mechanism was supported

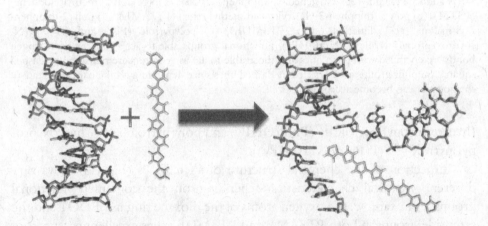

Figure 2: Mechanism proposed in reference 26 for the interaction between plasmid DNA and ECPs. This mechanism, which is based on the intercalation of the polymer between the two strands of DNA, was derived from CD and UV-Vis spectroscopic results.

by the significant increase of the UV-Vis signals associated to the DNA bases, which indicates that such building blocks are less protected by the secondary structure (i.e. more accessible to the environment). However, the degree of accessibility depends on the number of hydrogen bonding interactions between the bases and the polar groups.[26] The proposed mechanism is also compatible with observations related with the formation of ECP nanowires by DNA-templating. According to these observations, the ECP, initially formed as short cationic oligomers, binds to the DNA through electrostatic plus additional non-covalent interactions in a manner analogous to small groove binding drug molecules.[28]

The existence of specific PEDOT· · · plasmid DNA hydrogen bonds was also proven by investigating the possible formation of adducts in mixtures involving ECP with different doping levels: 0.14, 0.54 and 1.05 positive charges per repeat unit (denoted PEDOT-red, PEDOT-0 and PEDOT-ox, respectively).[29] PEDOT-0 and PEDOT-ox were found to form stable adducts and specific interactions with plasmid DNA, whereas PEDOT-red only interacted with DNA when the concentration of ECP was very high. These results suggested that non-specific electrostatic interactions between negatively charged DNA molecules and positively charged PEDOT molecules are essential to form stable adducts, the stability of complexes increasing with the doping level. Once adducts are stabilized weak interactions that depend on the spatial disposition and orientation of the chemical groups are formed. The formation of these specific interactions has been associated with the previously discussed accessibility of DNA bases (Fig. 2). Thus, the microscopic interpretation of the spectroscopic and electrophoretic results was as follows: B-DNA arrangement of the plasmid DNA undergoes significant structural alterations when it inter-acts with positively charged PEDOT chains (i.e. PEDOT-0 and PEDOT-ox). In contrast, the formation of specific interactions in DNA:PEDOT-red mixtures was thought to be obstructed by the absence of electrostatic interactions: the exposition of the bases was low because structural alterations in DNA were small.

4. Theoretical Studies of Specific Interactions Between ECP and DNA Building Blocks

Nowadays, high level QM methods, which are able to describe small molecular systems very accurately, can be successfully applied to analyse the chemical nature and strength of the specific interactions formed between the repeat units and/or building blocks of different (bio)macromolecules.[30,31] A few years ago, we evaluated the ability of pyrrole (Py), thiophene (Th) and 3,4-ethylendioxythiophene (EDOT), which are the building blocks of the most common ECPs (Figure 1), to interact through specific hydrogen bonds with the methylated analogues of DNA bases[20,21]: 9-methyladenine (mA), 9-methylguanine (mG), 1-methylcytosine (mC), and 1-methylthymine (mT). In order to avoid contamination due to electrostatic effects typically displayed by charged species, Py, Th and EDOT building blocks were considered in the neutral (reduced) state rather in the doped (oxidized) state. This was a correct approximation since it is well known that in doped ECPs charges are not uniformly distributed along the whole molecular chain but localized in small segments containing a low number of repeat units.[32-34] According to the experimental results on PEDOT-DNA adducts described in the previous section, neutral blocks also participate in the formation of specific hydrogen-bonding interactions between doped polymer chains and DNA bases.[29]

Calculations on Py···mNA, EDOT···mNA and Th···mNA complexes (where NA = A, G, C and T) were performed at the *ab initio* MP2 theoretical level. Initial structures for these complexes were constructed considering that Py is a donor and acceptor of hydrogen bonds while Th and EDOT act as acceptors only. According to such considerations, 42, 32 and 47 geometries were prepared for the Py···mNA, Th···mNA and EDOT···mNA complexes, respectively (i.e. for each interaction side of each nucleic acid base, different orientations of the ECP building block were assumed).[20,21] All these geometries were optimized combining the MP2 theoretical method with the 6-31G(d) basis set, the resulting level being denoted MP2/6-31G(d). In addition, single point energy calculations were performed on the optimized geometries using the same *ab initio* procedure combined with the 6-311++G(d,p) basis set (i.e. MP2/6-311++G(d,p) level). The binding energy (ΔE_b) for each complex was derived from the following equation:

$$\Delta E_b = E_{complex} - E_{ECP} - E_{mNA} \qquad (4)$$

where $E_{complex}$ corresponds to the energy of the optimized complex, and E_{ECP} and E_{mNA} are the energies of the isolated ECP and DNA building blocks obtained using the geometry of the complex. The basis set superposition error was corrected for each complex using the counterpoise method.[35]

Figure 3 displays the most stable structure found for each family of Py···mNA, and EDOT···mNA complexes, while the corresponding ΔE_b values are listed in Table 1. The number of relevant minimum energy structures identified for each complex ($N_\#$), which is defined as the number of local minima destabilized by less than 1.5 kcal/mol with respect to the most stable

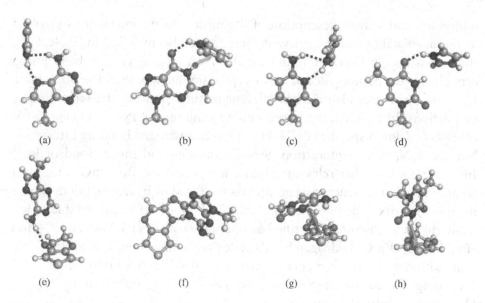

Figure 3: Most stable structure identified for: (a) Py$\cdots$mA; (b) Py$\cdots$mG; (c) Py$\cdots$mC; (d) Py$\cdots$mT; (e) EDOT$\cdots$mA; (f) EDOT$\cdots$mG; (g) EDOT$\cdots$mC; (h) EDOT$\cdots$mT.

Table 1: Binding energy for the lowest energy minimum (ΔE_b), number of relevant minima ($N_\#$) and description of the interactions in the lowest energy and relevant minima for Py$\cdots$mNA, EDOT$\cdots$mNA and Th$\cdots$mNA complexes.

	ΔE_b (kcal/mol)	$N_\#$	
		Py$\cdots$mNA	
Py$\cdots$mA	-9.3	3	(Py)N–H$\cdots$N(mA), (mA)N–H$\cdots$N(Py)
Py$\cdots$mG	-12.5	0	(Py)N–H$\cdots$O(mG), (mG)N–H$\cdots\pi$ (Py)
Py$\cdots$mC	-9.7	2	(Py)N–H$\cdots$N(mC), (mC)N–H$\cdots$N(Py)
Py$\cdots$mT	-7.7	4	
		EDOT$\cdots$mNA	
EDOT$\cdots$mA	-6.9	3	(mA)N–H$\cdots$O(EDOT), (mA)N–H$\cdots\pi$(EDOT)
EDOT$\cdots$mG	-9.8	4	(mG)N–H$\cdots$O(EDOT), (mG)N–H$\cdots\pi$(EDOT)
EDOT$\cdots$mC	-6.0	0	$\pi\cdots\pi$ stacking
EDOT$\cdots$mT	-10.3	2	$\pi\cdots\pi$ stacking, (mT)N–H$\cdots$O(EDOT)
		Th$\cdots$mNA	
Th$\cdots$mA	-4.5	2	(mA)N–H$\cdots\pi$(Th)
Th$\cdots$mG	-6.8	1	(mG)N–H$\cdots\pi$(Th)
Th$\cdots$mC	-4.9	0	$\pi\cdots\pi$ stacking
Th$\cdots$mT	-3.8	0	(mT)N–H$\cdots\pi$(Th)

minimum, and a short description of the most relevant interactions involved in the most stable and the relevant minima are also included in Table 1. In the lowest energy structure found for Py$\cdots$mA (Fig. 3(a)), the N–H group acts a hydrogen bonding donor and acceptor simultaneously,[21] the other three Py$\cdots$mA structures identified as relevant minima showing the same kind of interactions. In contrast, the lowest energy minimum of Py$\cdots$mG (Fig. 3(b)) exhibits not only a specific (Py)N–H$\cdots$O(mG) hydrogen bonding interaction but also a N–H$\cdots\pi$ interaction between the mG and the π-cloud of Py.[21] Interestingly, no other relevant minima was found for Py$\cdots$mG. The most favoured Py$\cdots$mC structure (Fig. 3(c)) is stabilized by hydrogen bonds similar to those described for Py$\cdots$mA (i.e. the N–H moiety of Py plays a dual role), while the two structures identified as relevant show (Py)N–H$\cdots$N(mC) and (Py)N–H$\cdots$O(mC) hydrogen bonds. The five minima identified for Py$\cdots$mT were within a relative free energy interval of 0.7 kcal/mol only, all of them (including the lowest energy one, Fig. 3(d)) being stabilized by (Py)N–H$\cdots$O(mT) hydrogen bonds.

The lowest energy minimum predicted for EDOT$\cdots$mA (Fig. 3(e)) exhibits a strong hydrogen bond between the amino group of mA and one of the oxygen atoms of the dioxane ring.[20] Furthermore, three local minima stabilized by (mA)N–H$\cdots$O(EDOT) or (mA)N–H$\cdots\pi$(EDOT) interactions were identified within a relative free energy interval $\leq$1.5 kcal/mol. Similarly, the lowest energy minimum identified for EDOT$\cdots$mG is stabilized by a (mG)N–H$\cdots$O(EDOT) hydrogen bond (Fig. 3(f)), while the four representative local minima displayed (mG)N–H$\cdots$O(EDOT) or (mG)N–H$\cdots\pi$(EDOT) interactions.[20] In opposition, the lowest energy minimum found for both EDOT$\cdots$mC and EDOT$\cdots$mT do not show hydrogen bonding interactions but π-π stacking (Figs. 3(g) and 3(h), respectively).[20] No other relevant minimum was identified for EDOT$\cdots$mC, whereas EDOT$\cdots$mT showed two local minima very close in energy to the global minimum and stabilized by (mT)N–H$\cdots$O(EDOT) hydrogen bonds.

Comparison of the ΔE_b calculated for the lowest energy minimum of Py$\cdots$mNA and EDOT$\cdots$mNA (Table 1) reveals the following features, which are important from a recognition point of view:

(i) Specific hydrogen bonds play a crucial role in the recognition of DNA bases by ECPs building blocks.

(ii) The affinity towards DNA bases is completely different for Py and EDOT, which should be attributed to the chemical characteristics of the oxygen atoms of EDOT (acceptor of hydrogen bonds) and the N–H group of Py (donor and acceptor of hydrogen bonds).

(iii) The strength of the binding for EDOT and Py complexes grows in the following order: mA < mC < mG ≈ mT and mT < mA ≈ mC < mG, respectively.

(iv) The very high affinity of Py and EDOT towards mG is enthalpic and entropic (0 and 4 relevant local minima were identified), respectively.

The success of these theoretical predictions was proved in a recent work,[36] in which UV-vis spectroscopy and QM calculations were used to examine the specific interactions in complexes formed between doubly protonated guanine (GH_2^{2+}) and EDOT. Thus, the acidic pH required to prepare EDOT:base mixtures with mass ratios of 1:1, 1:2, 2:1 and 1:4 induced the protonation of guanine that transformed into GH_2^{2+}. In spite of this, EDOT $\cdots GH_2^{2+}$ was considered as a reliable model complex since the coexistence of negatively charged groups in DNA and the positive charges of oxidized ECPs was avoided.

DFT calculations indicated that EDOT $\cdots GH_2^{2+}$ complexes are stabilized by N–H $\cdots$ O interactions involving an EDOT oxygen and the −NH and −NH$_2$ moieties of GH_2^{2+} (Fig. 4(a)).[36] Furthermore, time-dependent DFT calculations reproduced the adsorption spectra (both energy gaps and relative oscillator strength magnitudes), not only for EDOT and GH_2^{2+} but also for the complex (Fig. 4(b)).[36] Overall, those results provided clear evidence for the existence of specific interactions between PEDOT and DNA building blocks, allowing confirmation of previous mechanistic hypothesis derived from experimental observations.

On the other hand, Gao et al.[37] prepared porous films of overoxidized PPy/graphene nanocomposite that were successfully used for the quantitative detection of adenine and guanine. Due to the negative charge and structure of such nanocomposite, the detection of these bases was attributed to the formation of strong π-π interactions and electrostatic adsorption rather than to specific hydrogen bonds. Nevertheless, it should be mentioned that the recognition mechanism was hypothesized without any experimental or theoretical evidence, with no explicit investigation about the nature of the nanocomposite $\cdots$ bases interactions.[37]

The spectrum of diluted EDOT presented two important transitions at 259 and 266 nm while the GH_2^{2+} transitions were observed at 249 and 276 nm.[36] For the complex, the deconvolution process led to three peaks centred on 247, 257, and 281 nm. Accordingly, the interactions between the two species, responsible for the formation of the complex, considerably affected one of the transitions. Although the peaks at 259 and 249 nm for EDOT and GH_2^{2+}, respectively, underwent a change upon complexation (i.e. a 2 nm blue

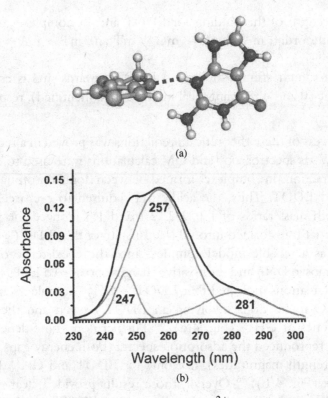

(a)

(b)

Figure 4: (a) Molecular structure of the EDOT$\cdots$GH$_2^{2+}$ complex derived from DFT QM calculations. (b) Absorption spectrum (black line) recorded for the EDOT$\cdots$GH$_2^{2+}$ complex (1:1 mass ratio) with the curves resulting from the deconvolution process spectra (gray lines).

shift was detected), the peaks at 266 and 276 nm merged and red shifted to 281 nm.

Figure 5 shows the most stable complex predicted by *ab initio* QM calculations for each Th$\cdots$mNA complex, the corresponding ΔE_b and $N_\#$ values being included in Table 1. The molecular geometry of the three Th$\cdots$mA minima (i.e. the lowest energy minimum, which is depicted in Fig. 5(a), and two local minima unfavored by less than 1 kcal/mol) are stabilized by a N–H$\cdots\pi$ interaction between the π-cloud of the Th ring and the exocyclic –NH$_2$ group of mA.[21] The two Th$\cdots$mG complexes of lowest energy, which are isoenergetic (Fig. 5(b)) and only differ in the relative orientation of the Th ring with respect to the base, are stabilized by two co-existing (mG)N–H$\cdots\pi$(Th) interactions each one.[21] Regarding to Th$\cdots$mC, the two building blocks interact through $\pi\cdots\pi$ in the lowest energy structure (Fig. 5(c)). Local minima stabilized by N–H$\cdots\pi$ and hydrogen bonding interactions were also detected, even though they were not relevant

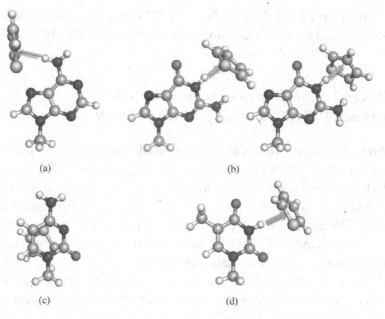

Figure 5: Most stable structure identified for: (a) Th$\cdots$mA; (b) Th$\cdots$mG (the two depicted structures, which are isoenergetic, only differ in the relative orientation of the two building blocks); (c) Th$\cdots$mC; (d) Th$\cdots$mT.

(i.e. such minima were disfavoured by more than 2 kcal/mol with respect to the lowest energy minimum). Finally, the only minimum identified for the Th$\cdots$mT complex (Fig. 5(d)) is stabilized by a N–H$\cdots\pi$ interaction.[21]

The values of ΔE_b obtained for Th$\cdots$mNA (Table 1) indicates that the strength of intermolecular interactions varies in the following order: Th$\cdots$mT < Th$\cdots$mA < Th$\cdots$mC < mT$\cdots$mG, which is identical to that of Py$\cdots$mNA complexes. However, the ΔE_b are significantly lower for Py$\cdots$mNA and EDOT$\cdots$mNA than for Th$\cdots$mNA, indicating that DNA bases prefer interaction with PPy and PEDOT. This feature is fully consistent with the experimental results described in the previous section, corroborating the fundamental role of specific hydrogen bonding interactions. Thus, the affinity of PTh derivatives (e.g. P3TM) toward plasmid DNA is significantly lower than that of PPy and PEDOT, even though the doping level of the different ECPs was relatively similar.[20,25,27] Specifically, gel electrophoresis assays performed for a series of PPy-DNA, PEDOT-DNA and P3TM-DNA complexes considering different polymer-DNA mass ratios showed the formation of complexes for 1:1, 1:1 and 100:1 mass ratios, respectively. The strength of electrostatic interactions between the ECP and the DNA is expected to be proportional to the doping level of the former, while hydrophobic interactions are expected to be stronger for P3TM than for PPy and PEDOT. Accordingly, the different affinities observed for such ECPs are probably related to the ability of PPy and PEDOT to form hydrogen bonds.

5.   Theoretical Studies of Specific Interactions Between PEDOT and Single Strand DNA

Computer simulations studies using classical MD simulations at the atomistic level on an oxidized PEDOT molecule interacting with the well-known Dickerson's dodecamer sequence[38] provided relevant information about the structure, stability and dynamics of PEDOT···DNA complexes.[39] The sequence of the Dickerson's dodecamer, which exhibits CG, AA and TT primary characteristic tracts, is 5′-CGCGAATTCGCG-3′. According to the mechanism proposed for the intercalation of the ECP chain between the two strands of double-helix DNA (Figure 2), the single strand was considered for such computational study. MD simulations were performed considering four specific arrangements as starting structures (Fig. 6(a)):

(A) The DNA strand maintains the same conformation that in a double helix while the PEDOT molecule faces the phosphate groups of the DNA chain.
(B) The DNA strand maintains the same conformation in a double helix while the PEDOT molecule faces the bases of the DNA chain.
(C) The DNA strand adopts a coiled conformation wrapping the PEDOT molecule.
(D) The DNA strand adopts a coiled conformation facing the PEDOT molecule.

On the other hand, the PEDOT chain was represented using an oligomer of 20 repeating units with a positive charge every two repeat units (i.e. +0.5 per EDOT unit) and 10 ClO_4^- as counterions. It should be noted that these conditions correspond to those employed to study the interaction of PEDOT and plasmid DNA,[20] the experimental doping level of this ECP synthesized by potentiometric methods with $LiClO_4$ as supporting electrolyte being +0.549.[40]

Independent MD simulations of the four starting arrangements A–D, converged in only two distinguishable models (Fig. 6(b)), which can be described as:

(C′) The DNA strand, which presents a coiled conformation, wraps the PEDOT molecule.

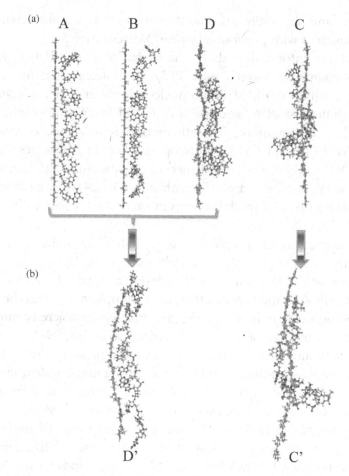

Figure 6:   (a) Representation of the PEDOT and DNA chains in the four starting structures used to study the complexation of such ECP with a Dickerson's dodecamer strand. (b) Models C' and D' obtained after independent MD simulations of A–D. Structures A, B and D converged into D' while C retained its main characteristics in C'.

(D') The PEDOT molecule and the coiled DNA conformation are faced side-by-side.

Detailed analysis indicated that, in addition of the electrostatic interactions between the negatively charged phosphate groups of DNA and the positively charged repeat units of PEDOT, specific hydrogen bonds and $\pi \cdots \pi$ stacking interactions were formed.[39] Although N–H$\cdots$O were very abundant and exhibited relatively large accumulated life times, $\pi \cdots \pi$ interactions were the most frequent. In terms of occurrence, specific interactions were more frequent with thymine and, especially, guanine than with adenine and cytosine, which was fully consistent with previous *ab initio* QM results.[20,21]

To get more information about the preferences of PEDOT, MD simulations on complexes formed by a PEDOT molecule that interacts with a 6-mer of a single stranded homo-nucleotide of guanine, adenine, cytosine and thymine (ss-dG6, ss-dA6, ss-dC6 and ss-dT6, respectively) were performed.[41,42] Furthermore, these theoretical results were complemented with UV-Vis, FTIR and CD spectroscopy studies on complexes formed by mixing PEDOT and both 6- and 24-mers of single stranded homonucleotides (ss-dGn, ss-dAn, ss-dCn and ss-dTn with $n = 6$ or 24), as well as with DFT QM calculations on small model systems made of the corresponding building blocks.

Spectroscopic results indicated that ss-dG6 form stable and compact adducts with PEDOT, which are clearly dominated by specific hydrogen bonds.[41] Atomistic MD simulations considering a starting B-DNA conformation for the ss-dG6 strand revealed that, upon complexation with the PEDOT chain, the homo-nucleotide unfolds into a completely disordered conformation allowing the formation of N–H···O hydrogen bonds, N–H···π, π···π and electrostatic interactions (Fig. 7(a)). In spite of such variety, hydrogen bonds were the most abundant and relevant among non-covalent interactions in PEDOT···ss-dG6 complexes. Overall, these figures were fully consistent with results obtained for the Dickerson's dodecamer.[39] Moreover, energy decomposition analysis (EDA),[43–46] which was performed by means of QM calculations on model compounds, showed that the most stabilizing interaction between the building blocks of this complex corresponds to the N–H···O hydrogen bonds. Thus, the formation of such kind of specific interaction between EDOT and guanine was found to increase the importance of resonance form B (Scheme 1) in guanine and, therefore, in this case N–H···O interactions were considered as resonance assisted hydrogen bonds.[47]

On the other hand, spectroscopic results suggested that ss-dAn and, especially, ss-dCn form adducts dominated by non-specific electrostatic interactions, while complexes with ss-dTn displays a behaviour different from those of adenine- and cytosine-homonucleotides.[42] Results obtained for ss-dTn with $n = 6$ and 24 suggests some resemblances to guanine-containing systems, reflecting certain specificity when interacting with PEDOT. However, the ability of ss-dTn to retain the association with PEDOT proved to be weaker than that observed in guanine-homonucleotides.

Results obtained from MD simulations were in agreement with such experimental observations. Thus, ss-dA6 was found to bind PEDOT preferentially

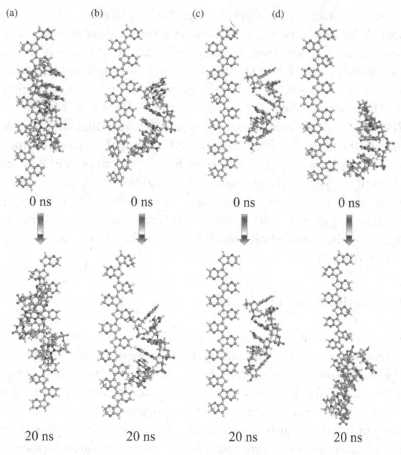

Figure 7: Initial and final snapshots from the MD simulations of: (a) PEDOT···ss-dG6 (b) PEDOT···ss-dA6 (c) PEDOT···ss-dC6 and (d) PEDOT···ss-dT6, respectively.

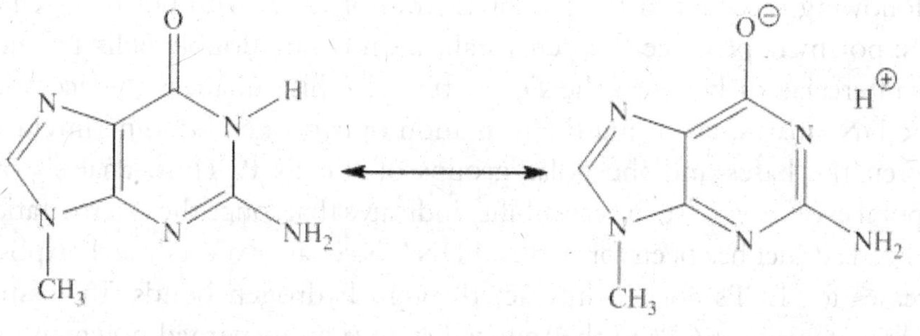

Scheme 1: Representative resonance forms of 1-methylguanine.

through $\pi \cdots \pi$ stacking interactions (Fig. 7(b)). The detected PEDOT $\cdots$ ss-dA6 specific hydrogen bonds were very unstable, evident by their very short residence times.[42] These features indicated that the association of adenine-homonucleotides with PEDOT was dominated by electrostatic interactions. The behaviour of ss-dC6 was more radical, the electrostatic being only representative interactions detected throughout the whole MD simulations of PEDOT $\cdots$ ss-dC6 complexes (Fig. 7(c)).[42] Thus, although $\pi \cdots \pi$ were regularly formed, they quickly disappeared leaving the cytosine exposed to the solvent. On the other hand, complexes formed by PEDOT and ss-dT6 showed a dynamical geometric reorganization (Fig. 7(d)), which was essentially attributed to an improvement of the electrostatic interactions.[42] This feature suggested that, despite a certain tendency to form specific hydrogen bonds, thymine-oligonucleotides prefer electrostatic interactions over directional interactions.

6.   Conclusions and Outlook

This Chapter documents the interaction between ECPs and DNA, which has been examined using a combination of experimental and theoretical tools. The ability of positively charged molecules to interact with DNA is based on the propensity of the latter to form adducts with cationic species. However, the affinity of ECPs bearing groups able to act as donors and/or acceptors of hydrogen bonds towards specific nucleotide sequences, as those found for the target sequences of different restriction enzymes, requires the formation of directional and specific interactions like hydrogen bonds. This feature forms the basis for the conclusion that the formation of specific interactions with DNA depends on the chemical nature of the ECP. Spectroscopic results led to propose the following mechanism for the interaction of DNA with positively charged ECPs: polymers produce the denaturalization of the double helix promoting their intercalation between the DNA strands. This enhances the accessibility of the DNA bases, favouring the formation of hydrogen bonding interactions between the bases and the polar groups of the ECP. Thus, analysis of the temporal evolution of such accessibility indicates that, once the electrostatically-stabilized adduct has been formed and DNA bases are exposed, such exposition decreases for ECPs able to interact through hydrogen bonds. Interestingly, experiments using ECPs in the reduced state (i.e. uncharged polymers) show evidence for the formation of specific interactions with DNA is obstructed by the absence of electrostatic interactions. In other words, in such cases, structural alterations in DNA are small and, therefore, the initial exposition of the DNA bases is low.

QM calculations show that the interactions with DNA building blocks are stronger for PEDOT and PPy than for PTh repeat units. Indeed, EDOT and Py usually interact through specific hydrogen bonds, while Th only forms complexes stabilized by interactions between the π-cloud of the ring and the N–H groups of nucleic acid bases. These features are in excellent agreement with the experimental results obtained for adducts formed by plasmid DNA and PEDOT, PPy and P3TM. The affinity towards DNA bases are completely different for EDOT and Py, which can be attributed to the chemical characteristics of the oxygen atoms of EDOT (hydrogen bonding acceptor) and the N–H group of Py (hydrogen bonding donor and acceptor).

Spectroscopic studies on mixtures of PEDOT and different homonucleotides point towards three main tendencies depending on the nucleotide bases type: (i) adenine- and cytosine-homonucleotides essentially interact with PEDOT through non-specific interactions; (ii) complexes involving guanine-homonucleotides are dominated by specific hydrogen bonds; and (iii) T-homonucleotides show a behaviour intermediate between those of (i) and (ii), exhibiting certain specificity when interacting with PEDOT. Results derived from MD simulations were in complete agreement with such experimental observations. Thus, PEDOT forms $\pi \cdots \pi$ stacking and electrostatic interactions with ss-dA6 and ss-dC6, respectively, while the interaction with ss-dG6 occurs via N–H$\cdots$O hydrogen bonds.

Overall, results discussed in this Chapter indicate that ECPs bearing polar groups able to interact through hydrogen bonds, which are typically biocompatible materials, are potential candidates for the development of electroactive devices able to act through specific molecular recognition patterns. Furthermore, this information is also useful for the development of new ECPs able to deliver drugs at specific regions of DNA.

Acknowledgements

The present work was financially supported by MINECO (MAT2015-69367-R) and the Generalitat de Catalunya (XRQTC). Authors are in debt to BSC (BCV-2009-2-0004) and CESCA for computational facilities. C.A. is grateful to ICREA Academia program.

References

1. C. Li, C. Bai and G. Shi, Conducting polymer nanomaterials: Electrosynthesis and applications. *Chem. Soc. Rev.*, **38**, 2397–2409 (2009).
2. D. W. Hatchett, M. Josowicz, Composites of intrinsically conducting polymers as sensing nanomaterials. *Chem. Rev.*, **108**, 746–769 (2008).

3. S. Geetha, C. R. K. Rao, M. Vijayan, D. C. Trivedi, Biosensing and drug delivery by polypyrrole. *Anal. Chim. Acta*, **568**, 119–125 (2006).

4. V. C. Ferreira, A. I. Melato, A. F. Silva, L. M. Abrantes, Conducting polymers with attached platinum nanoparticles towards the development of DNA biosensors. *Electrochem. Commun.*, **13**, 993–996 (2011).

5. L. J. del Valle, F. Estrany, E. Armelin, R. Oliver, C. Alemán, Cellular adhesion, proliferation and viability on conducting polymer substrates. *Macromol. Biosci.*, **8**, 1144–1151 (2008).

6. N. K. Guimard, N. Gomez, C. E. Schmidt, Conducting polymers in biomedical engineering. *Prog. Polym. Sci.*, **32**, 876–921 (2007).

7. A.-D. Bendrea, L. Cianga, I. Cianga, Progress in the field of conducting polymers for tissue engineering applications. *J. Biomater. Appl.*, **26**, 3–84 (2011).

8. R. Racichandran, S. Sundarrajan, J. R. Venugopal, S. Mukherjee, S. Ramakrishna, Applications of conducting polymers and their issues in biomedical engineering. *J. R. Soc. Interface*, **7**, S559–S579 (2010).

9. J. Jaguar-Grodzinski, Biomedical applications of electrically conductive polymeric systems. *e-Polymers*, **12**, 722–740 (2012).

10. C. Rincón, J. C. Meredith, Osteoblast adhesion and proliferation on poly(3-octylthiophene) thin films. *Macromol. Biosci.*, **10**, 258–264 (2010).

11. M. M. Pérez-Madrigal, E. Armelin, J. Puiggalí, C. Alemán, Insulating and semiconducting polymeric free-standing nanomembranes with biomedical applications. *J. Mater. Chem. B*, **3**, 5904–5932 (2015).

12. T. F. Otero, J. G. Martínez, J. Arias-Pardilla, Biomimetic electrochemistry from conducting polymers. A review: Artificial muscles, smart membranes, smart drug delivery and computer/neuron interfaces. *Electrochim. Acta*, **84**, 112–128 (2012).

13. M. Ates, A review study of (bio)sensor systems based on conducting polymers. *Mater. Sci. Eng. C*, **33**, 1853–1859 (2013).

14. G. Fabregat, E. Córdova-Mateo, E. Armelin, O. Bertran, C. Alemán, Ultrathin films of polypyrrole derivatives for dopamine detection. *J. Phys. Chem. C*, **115**, 14933–14941 (2011).

15. E. Córdova-Mateo, J. Poater, B. Teixeira-Dias, O. Bertran, F. Estrany, L. J. del Valle, M. Solà, C. Alemán, Electroactive polymers for the detection of morphine. *J. Polym. Res.*, **21**, 565 (2014).

16. G. Fabregat, J. Casanovas, E. Redondo, E. Armelin, C. Alemán, A rational design for the selective detection of dopamine using conducting polymers. *Phys. Chem. Chem. Phys.*, **16**, 7850–7861 (2014).

17. J. Casanovas, J. Preat, D. Zanuy, C. Aleman, Sensing abilities of crown ether functionalized polythiophenes. *Chem. Eur. J.*, **15**, 4676–4684 (2009).

18. D. Zanuy, J. Preat, E. A. Perpète, C. Alemán, Response of crown ether functionalized polythiophenes to alkaline ions. *J. Phys. Chem. B*, **116**, 4575–4583 (2012).

19. D. Aradilla, F. Estrany, C. Aleman, Polypyrrole derivatives as solvent vapor sensors. *RSC Adv.*, **3**, 3, 20545–20558 (2013).

20. C. Aleman, B. Teixeira-Dias, D. Zanuy, F. Estrany, E. Armelin, L. J. del Valle, A comprehensive study of the interactions between DNA and poly(3,4-ethylenedioxythiophene). *Polymer*, **50**, 1965–1974 (2009).

21. D. Zanuy, C. Alemán, DNA-conducting polymer complexes: A computational study of the hydrogen bond between building blocks. *J. Phys. Chem. B*, **112**, 3222–3230 (2008).

22. E. Lewars, Computational Chemistry: Introduction to the Theory and Applications of Molecular and Quantum Mechanics. Kluwer Academic Publishers, Dordrecht (2004).

23. F. Jensen, *Introduction to Computational Chemistry*. 2nd ed, John Wiley & Sons Ldt, Chichester (2007).

24. A. Szabo, N. S. Ostlund, *Modern Quantum Chemistry: Introduction to Advanced Electronic Structure Theory*. Macmillian Publishing Co., Inc., New York (1982).

25. C. Ocampo, E. Armelin, F. Estrany, L. J. del Valle, R. Oliver, F. Sepulcre, C. Alemán, Copolymers of 3,4-ethylenedioxythiophene and 3-methylthiophene: Properties, applications and morphologies. *Macromol. Mater. Eng.*, **292**, 85–94 (2007).

26. B. Teixeira-Dias, L. J. del Valle, F. Estrany, E. Armelin, R. Oliver, C. Alemán, Specific interactions in complexes formed by polythiophene derivatives bearing polar side groups and plasmid DNA, *Eur. Polym. J.*, **44**, 3700–3707 (2008).

27. P. Pfeiffer, E. Armelin, F. Estrany, L. J. del Valle, L. Y. Chao, C. Alemán, Copolymers of pyrrole and N-(hydroxypropyl)pyrrole: Properties and interaction with DNA. *J. Polym. Res.*, **15**, 225–234 (2008).

28. S. M. D. Watson, M. A. Galindo, R. Horrocks, A. Houlton, Mechanism of formation of supramolecular DNA-templated polymer nanowires. *J. Am. Chem. Soc.*, **136**, 6649–6655 (2014).

29. B. Teixeira-Dias, D. Zanuy, L. J. del Valle, F. Estrany, E. Armelin, C. Alemán, Influence of the doping level on the interactions between poly(3,4-ethylenedioxythiophene) and plasmid DNA. *Macromol. Chem. Phys.*, **211**, 1117–1126 (2010).

30. I. A. W. Filot, A. R. A. Palmans, P. A. J. Hilbers, R. A. van Santen, E. A. Pidko and T. F. A. de Greef, Understanding cooperativity in hydrogen-bond-induced supramolecular polymerization: A density functional theory study. *J. Phys. Chem. B*, **114**, 13667–13674 (2010).

31. A. Jain, R. N. V. Deepak, R. Sankararamakrishnan, Oxygen-aromatic contacts in intrastrand base pairs: analysis of high-resolution DNA crystal structures and quantum chemical calculations. *J. Struct. Biol.*, **187**, 49–57 (2014).

32. J. Casanovas, C. Alemán, Comparative theoretical study of heterocyclic conducting polymers: Neutral and oxidized forms. C. *J. Phys. Chem. C*, **111**, 4823–4830 (2007).

33. J. L. Brédas, Relationship between band gap and bond length alternation in organic conjugated polymers. *J. Chem. Phys.*, **82**, 3808–3811 (1985).

34. V. Hernandez, C. Castiglioni, M. Del Zopo, G. Zerbi, Confinement potential and π-electron delocalization in polyconjugated organic materials. *Phys. Rev. B: Condens. Matter Mater. Phys.*, **50**, 9815–9823 (1994).

35. S. F. Boys, F. Bernardi, The calculation of small molecular interactions by the differences of separate total energies. Some procedures with reduced errors. *Mol. Phys.*, **19**, 553–566 (1970).

36. J. Preat, B. Teixeira-Dias, C. Michaux, E. A. Perpète, C. Alemán, Specific interactions in complexes formed by DNA and conducting polymer building blocks: Guanine and 3,4-(ethylenedioxy)thiophene. *J. Phys. Chem. A*, **115**, 13642–13648 (2011).

37. Y.-S. Gao, J.-K. Xu, L.-M. Lu, L.-P. Wu, K.-X. Zhang, T. Nie, X.-F. Zhu, Y. Wu, Overoxidized polypyrrole/graphene nanocomposite with good electrochemical performance as novel electrode material for the detection of adenine and guanine. *Biosens. Bioelectron.*, **62**, 261–267 (2014).

38. H. R. Drew, R. M. Wing, T. Takano, C. Broka, S. Tanaka, K. Itakura, R. E. Dickerson, Structure of a B-DNA dodecamer: Conformation and dynamics. *Proc. Natl. Acad. Sci. U.S.A.*, **78**, 2179–2183 (1981).

39. J. Preat, D. Zanuy, E. A. Perpetè, C. Alemán, Binding of cationic conjugated polymers to DNA: Atomistic simulations of adducts involving the Dickerson's dodecamer. *Biomacromolecules*, **12**, 1298–1304 (2011).

40. C. Ocampo, R. Oliver, E. Armelin, C. Alemán, F. Estrany, Electrochemical synthesis on steel electrodes of poly(3,4-ethylenedioxythiophene): Properties and characterization. *J. Polym. Res.*, **13**, 193–200 (2006).

41. B. Teixeira-Dias, D. Zanuy, J. Poater, M. Solà, F. Estrany, L. J. del Valle and C. Alemán, Binding of 6-mer single-stranded homo-nucleotides to poly(3,4-ethylenedioxythiophene): Specific hydrogen bonds with guanine. *Soft Matter*, **7**, 9922–9932 (2011).

42. D. Zanuy, B. Teixeira-Dias, L. J. del Valle, J. Poater, M. Solà, C. Alemán, Examining the specific interactions between poly(3,4-ethylenedioxythiophene) and nucleotide bases. *RSC Adv.*, **3**, 2639–2649 (2013).

43. K. Morokuma, Why do molecules interact? The origin of electron donor-acceptor complexes, hydrogen bonding and proton affinity. *Acc. Chem. Res.*, **10**, 294–300 (1977).

44. K. Kitaura, K. Morokuma, A new energy decomposition scheme for molecular interactions within the Hartree-Fock approximation. *Int. J. Quantum Chem.*, **10**, 325–340 (1976).

45. T. Ziegler, A. Rauk, On the calculation of bonding energies by the Hartree Fock Slater method. *Theor. Chim. Acta*, **46**, 1–10 (1977).

46. T. Ziegler, A. Rauk, Carbon monoxide, carbon monosulfide, molecular nitrogen, phosphorus trifluoride, and methyl isocyanide as .sigma. donors and .pi. acceptors. A theoretical study by the Hartree-Fock-Slater transition-state method. *Inorg. Chem.*, **18**, 1755–1759 (1979).

47. M. Palusiak, S. Simon, M. Solà, Interplay between intramolecular resonance-assisted hydrogen bonding and aromaticity in o-hydroxyaryl aldehydes. *J. Org. Chem.*, **71**, 5241–5248 (2006).

Chapter 5

Supramolecular Assemblies of DNA/Conjugated Polymers

Jérémie Knoops,* Jenifer Rubio-Magnieto,*
Sébastien Richeter,† Sébastien Clément,†
and Mathieu Surin*,†,‡

*Laboratory for Chemistry of Novel Materials,
Center for Innovation in Materials and Polymers,
University of Mons — UMONS, 20 Place du Parc,
B-7000 Mons, Belgium

†Institut Charles Gerhardt — UMR 5253,
Université de Montpellier — CC1701,
Place Eugène Bataillon,
F-34095 Montpellier Cedex 05, France

Water-soluble π-conjugated polymers offer new perspectives as nanomaterials for biosensing and biomedical imaging. A series of cationic conjugated polymers were studied for DNA optical detection, for example to detect DNA lesions or single-nucleotide polymorphisms (SNPs), which is of great interest for genomic applications. The fundamental understanding of structural aspects of DNA/conjugated polymers complexes in aqueous and physiological solutions is key for the development of relevant biosensors. This chapter aims at providing a new perspective on DNA/conjugated polymer assemblies at the supramolecular level, by deciphering the effects of DNA sequence and related conformational changes and aggregation processes occurring upon the self-assembly of DNA with cationic polythiophenes. By combining (chir)optical spectroscopy and molecular modelling simulations, we provide helpful structural insights in the context of DNA hybridization biosensors, with an example applied to the detection of single nucleotide mismatch.

---

‡Corresponding author: mathieu.surin@umons.ac.be

1. Introduction

Water-soluble π-conjugated polymers, often referred to as π-conjugated polyelectrolytes (CPEs), have emerged as a promising class of nanomaterials for developments in biodetection and in biomedical imaging, owing to their remarkable properties as optical probes.[1-5] A particular effort has been devoted to the design of cationic CPEs for DNA binding through electrostatic self-assembly with the anionic phosphodiester backbone of DNA. Notable developments were carried out for determining DNA concentration in assays,[6] for detection of DNA hybridization mismatches and Single-Nucleotide Polymorphisms (SNPs),[7-10] and for regulating the gene expression.[11,12] Although the cationic CPE shows remarkable detection sensitivity, usually by exploiting fluorescence amplification (or quenching), the conformational changes of DNA and CPE upon their self-assembly, together with inter-polyelectrolyte aggregation effects, remain unclear. The effects of DNA sequence on the DNA/CPE structure have been scarcely studied,[13,14] which constitutes a strong limitation in the understanding of both selectivity and sensitivity of the proposed DNA biosensors. Besides, the development of DNA/CPE gene-delivery polyplexes requires a careful control over their supramolecular self-assembly, which has been overlooked so far.

As a matter of fact, studying the self-assembly of DNA/CPE at the molecular and supramolecular levels is of high importance for developing (bio)sensors and polyplexes. In this context, we and others have paid particular attention to the effects of DNA sequence, length, and topology, and the nature of the polymer cationic side groups on the supramolecular self-assembly.[13,15-17] Indeed, besides the electrostatic interactions, which constitute the main driving force for the self-assembly of DNA (anionic) with the cationic polymer, other types of supramolecular interactions are at play, such as π-π interactions between the aromatic monomers and the nucleobases, cation/anion — π interactions, and possibly H-bonding interactions, which influence the assembly and polymer conformational changes. Therefore, the DNA sequence strongly influences the optical detection properties[18,19] which can affect the homogeneity of the biosensor assays, as observed by Leclerc *et al.* for hybridization experiments with over 50 DNA sequences.[20]

In this chapter, we first introduce the systems under study, *i.e.* a series of polythiophenes with specific cationic side groups, in the context of DNA-based hybridization sensors (Section 2). In Section 3, we describe our studies on the supramolecular self-assembly of a selected cationic polythiophene with either a single-stranded DNA (ssDNA) or a double-stranded DNA (dsDNA). We particularly focus on the effects of DNA sequence. In Section 4, we use cationic polythiophenes as optical probes of DNA-DNA hybridization, it is to say to detect the formation the hybridized DNA duplex on a preformed ssDNA/cationic polythiophene complex. Finally, in Section 5 we give insights into the possible dsDNA/cationic polythiophene structures in view of specific interactions. The overall objective is shed light on the structural aspects of DNA/cationic polythiophene assemblies, which will be useful to develop CPE-based bioassays with DNA targets.

2. Cationic Polythiophenes for DNA Binding

A series of cationic π-conjugated polymers were designed to bind to DNA via electrostatic self-assembly with the anionic phosphodiester backbone. Among this series, the most studied cationic (co)polymers are derivatives of poly(2,7-fluorene)s (PFs),[21] poly(p-phenylene vinylene)s (PPVs),[14,19] and poly(thiophene)s (PTs).[1,2] Whereas PFs and PPVs show superior fluorescence quantum yields with respect to PTs, PTs were more frequently used for DNA hybridization sensing and as diagnostics reporters, because of the higher flexibility of the cationic polythiophenes (CPT) backbone, which allows for adapting to biomolecular targets with a large conformational diversity. This flexibility is an advantage for colorimetric assays, since planarization and (un)folding effects can lead to large shifts of UV-Vis absorption spectra (see below). Besides, CPT were also used in fluorimetric assays, because of the molecular wire effect that renders them suitable in fluorescence resonance energy transfer (FRET) assays, using for instance a DNA end-labelled dye as acceptor.

Owing to the availability of robust synthetic protocols (e.g. Kumada Catalyst-Transfer Polycondensation), polythiophene of multiple topologies (homopolymers, random/block copolymers) can be readily achieved with a high degree of control over the final composition and molecular weight.[22–24] In particular, we have reported the synthesis of well-defined poly(3-bromohexyl)thiophene polymers (and copolymers) with low dispersities (<1.4) and number-averaged molecular weights (M_n) between 13500 and 15000 g.mol^{-1}.[17,25,26] These bromide precursors permitted the preparation of various types of cationic polythiophenes-based CPEs with different cationic side groups, such as imidazolium, pyridinium, ammonium and phosphonium groups (Fig. 1, left).[17,25–27]

When mixing cationic polythiophenes (or other CPE) with DNA in aqueous solutions, the size of the (inter)polyelectrolyte complexes is indeed critically dependent on solution conditions, and in particular the concentration

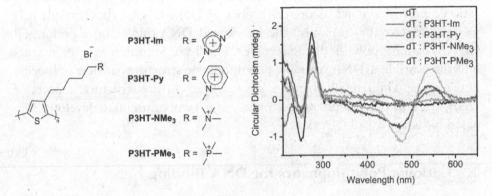

Figure 1:   Left: Chemical structure of the studied cationic polythiophenes with various cationic side groups studied. Right: circular dichroism spectra of mixtures of the various cationic polythiophenes with a single-stranded DNA (**dT**, see Chart 1). (Adapted from Ref. 17 with permission from The Royal Society of Chemistry.)

of each compound and the charge ratio.[15,28] In order to design DNA/cationic polythiophene complexes of interest for hybridization experiments, based on solubility and stability experiments we selected a polymer and DNA concentration of a few μM (typically 6–8 μM), a range of concentration where each compound is molecularly dissolved in aqueous solutions (using a Tris + EDTA buffer, pH = 7.4). In our conditions, the size of the DNA/CPT polyelectrolyte complexes is on the order of a few nm to few tens of nm, i.e. small polyelectrolyte complexes, considering that the typical size of bimolecular complexes is around 3–5 nm (as estimated with molecular modeling).[16]

Indeed, the nature of the cationic side group strongly influences the binding with DNA. We have shown that the size of this cationic group strongly influences the chirality of the assembly with DNA, as revealed by electronic circular dichroism spectroscopy (hereafter referred to as CD) in Fig. 1. For polymers with small cationic side groups (**P3HT-NMe₃** or **P3HT-PMe₃** for ammonium or phosphonium groups, respectively), upon DNA binding (with oligothymidine) chiral induction effects are manifested by CD signals in the range where the achiral polymers absorb light (420–580 nm), whereas pure cationic polythiophenes are CD silent in this range in aqueous solution.[17] However, for cationic polythiophenes with larger cations (**P3HT-Im** for imidazolium, **P3HT-Py** for pyridinium), very weak or no CD signal appear in this range. The intensity of the induced CD signals is correlated to the charge localization on the side groups, which could be exploited for chiroptical DNA sensing. Therefore, we have mainly focused our studies with **P3HT-PMe₃** (hereafter referred to as **CPT** for cationic polythiophene), which permit to relate chiroptical signals effects with the other usual spectroscopic signals such as UV-Vis absorption and fluorescence.

3. Effect of the DNA Sequence on the Self-assembly of CPT/DNA

To assess the effect of DNA sequence and topology (single-stranded *versus* double-stranded DNA) on the self-assembly with **CPT** (i.e., **P3HT-PMe₃** in Figure 1, M_n = 17700 g.mol^{-1}), we have selected well-defined DNA oligonucleotides (ODNs, shown in Chart 1), which contain a precise number of nucleobases, and represent model cases: oligothymidine (**dT**), which is a homopyrimidine ODN; oligoadenine (**dA**), a homopurine ODN; and mixed sequences with half purines and half pyrimidines (**dR, dP** and **dM**). The length of the oligonucleotides, around 20–25 base pairs (bp) in double-stranded configuration, was optimized to get approximately an equal number of DNA negative charges and polymer positive charges in 1:1 molar ratio

ssDNA:

dT₂₀: 5'-TTT TTT TTT TTT TTT TTT TT-3' (**dT**)

dA₂₀: 5'-AAA AAA AAA AAA AAA AAA AA-3' (**dA**)

dT₄₀: 5'-TTT TTT TTT TTT TTT TTT TTT TTT TTT TTT TTT TTT TTT TTT T-3'

dA₄₀: 5'-AAA AAA AAA AAA AAA AAA AAA AAA AAA AAA AAA AAA A-3'

dR₂₀: 5'-CGT CAC GTA AAT CGG TTA AC-3' (**dR**)

dR$_{rev20}$:   3'-GCA GTG CAT TTA GCC AAT TG-5' (**dR$_{rev}$**)

dR₄₃: 5'-CGT CAC GTA AAT CGG TTA ACA AAT GGC TTT CGA AGC TAG CTT C-3'

dP: 5'-AGG ATG GGC CTC <u>C</u>GG TTC ATG CCG C-3'

dM: 5'-AGG ATG GGC CTC <u>T</u>GG TTC ATG CCG C-3'

dP$_{rev}$: 5'-GCG GCA TGA ACC <u>G</u>GA GGC CCA TCC T-3'

dsDNA:

dA·dT: [5'-AAA AAA AAA AAA AAA AAA AA-3']
             [3'-TTT TTT TTT TTT TTT TTT TT-5']

dR·dR$_{rev}$:   [5'-CGT CAC GTA AAT CGG TTA AC-3']
             [3'-GCA GTG CAT TTA GCC AAT TG-5']

Chart 1:   Sequences of the selected DNA oligonucleotides (ODNs).

aqueous mixtures, based on results reported in Refs. 16 and 17. The formation of a supramolecular complex (complexation) upon electrostatic self-assembly between CPT with DNA in aqueous solution was studied by means of (chir)optical spectroscopy (UV-Vis absorption, circular dichroism, and fluorescence), which provide valuable information on conformational changes and aggregation processes.

The mixtures between single- or double-stranded ODNs (listed in Chart 1) and CPT were made in a tris-EDTA (TE) buffered solution MilliQ water at pH 7.4 and were equilibrated for 30 minutes. For all the DNA sequences, the complexation produces red-shifted UV-Vis absorption maxima in the region where the polymer absorbs compared to the pure CPT. This red-shift indicates the complex formation and the unfolding/planarization of the polymer chains upon DNA binding.[29–32] Importantly, at constant molar ratio, this red-shift varies with the DNA sequence. Mixtures with the homopurine dA_{20} oligonucleotide shows very large red-shift of the absorption maximum ($\Delta\lambda_{max} \sim +69$ nm) up to a 1:0.4 molar ratio in dA_{20}:CPT. Above this molar ratio, the shift gradually decreases, yielding at a 1:1 molar ratio a UV-Vis band, which shows a maximum intensity λ_{max} at 442 nm ($\Delta\lambda_{max} \sim -9$ nm) with a large shoulder in the 500–550 nm region. In contrast, for the homopyrimidine (dT_{20}) and for the mixed (dR_{20} and dR_{rev20}) oligonucleotides, the obtained $\Delta\lambda_{max}$ is around 42 nm at 1:1 molar ratio. These important differences in the λ_{max} for homopurine *vs* homopyrimidine or mixed sequences are likely arising from the strong binding between the CPT π-conjugated backbone and the aromatic plane of adenine (a purine), which is larger than thymidine (a pyrimidine). However, for mixtures of CPT with longer ssDNA sequences (dT_{40}, dA_{40} and dR_{43}), UV-Vis absorption spectra exhibit identical red-shifts, i.e. $\Delta\lambda_{max} \sim +46$ nm. Altogether, the shift of the main absorption band of CPT, related to the folding/unfolding of the polymer backbone, depends on the composition of the mixture. This is illustrated in Fig. 2 right, showing the results of a UV-Vis titration experiments. It shows that the shift of the maximum absorption wavelength (λ_{max}) of the polymer, red-shifted of around 55 nm at low concentrations of CPT in a CPT:dR mixture (with a DNA concentration of 7 μM), remains largely red-shifted up to a concentration of 3.2 μM in CPT. Above this concentration (i.e. above 1:0.45 in DNA:CPT molar ratio), the shift progressively decreases as a consequence of free CPT in the solution, because the 1:1 charge balance is attained at a molar ratio of around 0.44 in DNA:CPT (20 negative charges for the DNA and an average of 47 positive charges for the polymer, as estimated from Ref. 17 using a method described in Ref. 33). In order to visualize this red-shift, Fig. 2(b) shows the different colours of

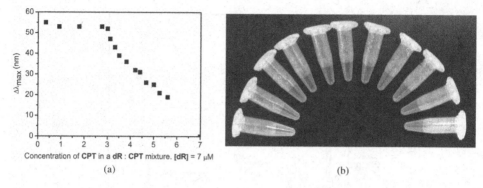

(a)　　　　　　　　　　　　　　　　　　(b)

Figure 2: (a) Shift of the **CPT** maximum absorption wavelength as a function of the concentration of **CPT** in the **dR** : **CPT** mixture (at a constant **dR** concentration of 7 μM). (b) Picture of **dR** : **CPT** mixtures at various compositions (total concentration $C_{dR} + C_{CPT} = 8.5$ μM, in TE buffer), going from a pure DNA solution on the left to a pure CPT solution on the right. (Adapted from Ref. 16 with permission from The Royal Society of Chemistry.)

dR$_{20}$/CPT complexes at various compositions of **dR$_{20}$:CPT** mixture, keeping constant the total concentration at around 8 μM.

The conformational changes of the polymer backbone upon binding to DNA strongly depend on the DNA topology, as observed when comparing mixtures with single-stranded DNA or double-stranded DNA. In the case of **CPT** self-assembly with a double helical DNA **dA·dT**, the polymer is less planar (more folded) than with corresponding single-stranded DNA, as indicated by the smaller red-shift of the absorption maximum of the polymer. Indeed, in dsDNA, the nucleobases are paired and base stacking (inter- and intra-strand) interactions are much more prevalent in duplexes than in single-strands. Therefore, the π-π interactions between the conjugated backbone and the nucleobases are unlikely.[18] In contrast, for the **dR·dR$_{rev}$/CPT** complex, the UV-Vis spectra show a similar red-shift of the absorption maximum ($\Delta\lambda_{max} \sim$ +42 nm) to that in **dR/CPT** (or **dR$_{rev}$/CPT**), which points out the effects of DNA sequence on polymer conformation.

CD spectroscopy can also be very interesting to understand the chiral induction effects of different DNA sequences upon complexation with the achiral **CPT**. We and others have observed that bisignate induced CD (ICD) signals signature appear in the wavelength range where the achiral polymer absorbs upon its complexation with the chiral DNA, which we attributed to a helical organization (conformation or aggregation) of the polymer chain in the DNA/polymer complex, as observed for other complexes made of biomolecules and conjugated oligomers or polymers.[34-36] We have observed that the DNA sequence strongly influences the CD signals of DNA/polymer

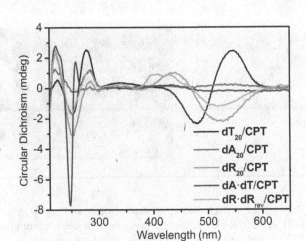

Figure 3:   Circular dichroism spectra of several ssDNA/CPT and dsDNA/CPT complexes in aqueous solution (at 1:1 molar ratio, 20°C TE buffer pH = 7.4).

complexes, with specific right-handed or left-handed signatures of the polymer within in the supramolecular complex. For instance, we observe a $(+/-)$ bisignate ICD signal in the region where the polymer absorbs, for binding with DNA made of homopyrimidine backbone (dT_{20} or dT_{40}), which indicates a preferential right-handed helical organisation of the polymer backbone within the dT/CPT complex;[17] whereas for the dA_{20}/CPT complex, there is no ICD signal in the polymer region, and there is a weaker $(+/-)$ bisignate ICD in the case of the complex with longer homopurine bases (dA_{40}), see Figure 3. This observation is in agreement with the UV-Vis absorption spectra, indicating a distortion of the helical conformation of the oligoadenine due to the strong π-π interactions between the purine bases and the aromatic monomers.[37] Interestingly, in the case of the mixed sequences (half pyrimidine and half purine bases), dR_{20}, dR_{rev20} or dR_{43} with **CPT**, the ICD observed is the signature of a preferential left-handed helical chirality of the polymer chain $((-/+)$ ICD), as shown in Figure 3.[16]

For complexes of **CPT** with preformed double-stranded DNA, the CD spectra also depend on the DNA sequences. Whereas complexes of **CPT** with $dA \cdot dT$ do not show any ICD effect, for the mixed sequences ODN studied ($dR \cdot dR_{rev}$), the CD spectra show a bisignate $(-/+)$ ICD signal, which suggests that the polymer is around the DNA helix in a preferential left-handed helical conformation or aggregation. Overall, these results show that the DNA sequence has major influence on the reorganization and chiral self-assembly of cationic polythiophenes with DNA, which could be used as a probe to detect specific oligonucleotides sequences.

4. DNA-DNA Hybridization Sensing Using CPT: Effect of Sequence

Given the high sensitivity of the (chir)optical signals of cationic polythiophenes to the DNA sequence when complexed, these signals could possibly be exploited for DNA-DNA hybridization detection. DNA-DNA hybridization may occur on a preformed ssDNA/CPT complex upon addition of a complementary DNA, ultimately forming so-called "triplexes".[1] To study the hybridization processes, fluorescence signals are commonly used, since the conformational changes of the polymer upon binding with the different DNA strands produce different fluorescence spectra. Many examples of label-free and dye-labeled DNA–CPE biosensors are reported in the literature, showing the great sensitivity of fluorescence signals of CPEs for the recognition of DNA targets.[10,19,20,29,31,38,39]

Again, the effect of DNA sequences on the complexes formed with the polymers is poorly understood. Based on the model sequences discussed in previous sections, we studied the hybridization processes of the following complexes: [**dT/CPT + dA**], [**dA/CPT + dT**], [**dR/CPT + dR$_{rev}$**] and [**dR$_{rev}$/CPT + dR**]. The fluorescence spectrum of **dA/CPT** complex at a 1:1 molar ratio is quenched compared to the fluorescence of the pure polymer at the same concentration (Fig. 4(a)), likely as a consequence of the polyelectrolyte aggregation, which can be favored by the strong interactions between the adenine nucleobases and the thiophene units.[16,40] Upon addition of the complementary strand **dT**, the fluorescence is partially recovered (Fig. 4(a)).[29,38] When the order of addition is reversed (**dT/CPT + dA**), we observed a higher fluorescence signal of the **dT/CPT** complex than

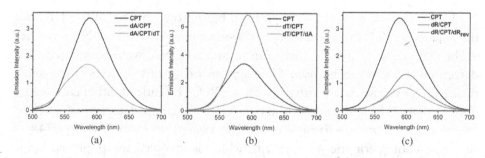

Figure 4: Emission spectra ($\lambda_{exc} = 450$ nm) of pure **CPT** (black line) and (a) **dA/CPT** (red line) and [**dA/CPT + dT**] (green line); (b) **dT/CPT** (red line) and [**dT/CPT + dA**] (green line); and (c) [**dR/CPT**] (red line) and [**dR/CPT + dR$_{rev}$**] (green line). Molar ratio are 1:1 for ssDNA/CPT and 1:1:1 for ssDNA/**CPT**/ssDNA$_c$ (TE buffer pH = 7.4).

for pure **CPT**, which can be explained by a conformation change of the polymer backbone from a coiled conformation when pure in aqueous solution (fluorescence self-quenching) to a more extended polymer conformation upon DNA complexation (Fig. 4(b)).[14,19] The addition of **dA** to this complex quenches the emission of the polymer. However, for mixed sequences the fluorescence is quenched upon addition of **dR** (or **dR$_{rev}$**) to **CPT** but not as much as in the case of **dA/CPT** (Fig. 4(c)). The fluorescence intensity of 'triplexes' is intermediate between the one obtained for the homobase sequences (in between [**dT/CPT + dA**] and [**dA/CPT + dT**]. Altogether, these results indicate that the fluorescence signals of **CPT** strongly depend on the DNA sequences in the triplex system, and also on the order of addition of the DNA sequences in the mixture.[20]

UV-Vis and CD spectra also show strong effects of the DNA sequences and the DNA strand addition in the hybridization experiments involving **CPT** in the complex. For instance, UV-Vis spectra of [**dT/CPT + dA**] show a blue-shifted maximum absorption, which is similar to the blue-shift observed for the complexation of **CPT** with a preformed double stranded **dA·dT**, suggesting again a lesser extent of planarization of the polymer upon binding with dsDNA compared to ssDNA. The induced CD signal diminishes, as a result of the non-planar conformation adopted by the polymer. In contrast, the CD spectra of [**dA/CPT + dT**] do not show any ICD signal in the range where the polymer absorbs. Notice that the UV-Vis maximum absorption band shifts to the blue in comparison with the one obtained for the [**dT/CPT + dA**], indicating aggregation of polymer chains in this system. Altogether, these various chiroptical signals show the effects of DNA sequence on the self-assembly with **CPT**, which yields different DNA/CPT aggregates. This can be further analyzed by melting temperature experiments on complexes, as detailed in Ref. 16.

In the context of DNA hybridization biosensors, the use of the **CPT** for the detection of SNPs, i.e. variations of a single nucleotide in a DNA sequence as the main cause of genetic variations in human DNA, was investigated. The detection of SNPs can be used for diagnosis of various diseases (for a short review on SNPs associated with diseases, see Ref. 41). Indeed, other conjugated polyelectrolytes have already proven to be effective for the detection of SNPs.[1,9] In our experiments, we focused on a SNP located on exon 7, codon 248 of the gene coding for the *p53* protein, which is involved in apoptosis. Some SNPs on this gene can perturb *p53*-mediated tumor suppression.[42,43] These ssDNA sequences have been used in our experiments: **dP**, **dM** and **dP$_{rev}$**, see sequence in Chart 1. The oligonucleotides **dP** and **dM** are the analytes;

and dP_{rev} oligonucleotide, complementary to dP, is used as probe. Note that those sequences have been used by Gerion *et al.*[44] to develop another type of SNP sensor using microarray-based detection. The solutions were prepared in the TE buffer at pH 8. dP (or dM) was first added to the buffer, then the CPT and finally the dP_{rev}, all of them from a stock solution. The final concentration for each DNA is 5 μM and the CPT final concentration is 2.5 μM (*i.e.* molar ratio 2:1 in DNA:CPT). Both solutions were heated and maintained at 66°C, which is approximately 5°C above the $dP \cdot dP_{rev}$ duplex melting temperature. Then, these solutions were progressively cooled down to 42.5°C in 20 minutes and finally to 20°C in 5 minutes. This protocol was developed to enable the hybridization of the two single-strands to form a double-stranded DNA, it is to say a Watson-Crick pairing of all bases between dP and dP_{rev} (homoduplex formation) or an incomplete base pairing between dM and dP_{rev} (heteroduplex formation due to the presence of a mismatch). The UV-Vis absorption spectra were subsequently recorded at 20°C. The absorption band of the CPT is red-shifted by 9 nm for the heteroduplex $[dM/CPT + dP_{rev}]$ compared to the homoduplex $[dP/CPT + dP_{rev}]$ (see Fig. 5). As shown above and also suggested by others,[1] this red-shift likely reflects the differences of polymer conformations between the two assemblies: in the presence of the DNA heteroduplex $dM \cdot dP_{rev}$, the CPT is more planar (more conjugated, such as it is with a single-stranded DNA) than in the presence of the homoduplex dP/dP_{rev}. The smaller red-shift in our example (compared to the results obtained by Ho *et al.*)[1] could be explained by a combination of factors, like the longer DNA length, the DNA sequence and the different chemical structure of the polymer side groups.

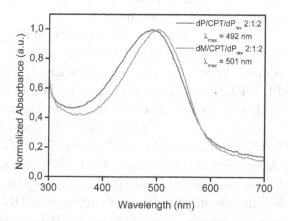

Figure 5: UV-Vis absorption spectra of hybridization experiments $[dP/CPT + dP_{rev}]$ and $[dM/CPT + dP_{rev}]$ (TE buffer, 20°C).

5. Supramolecular Organization into DNA/Conjugated Polymer Complexes

To understand the effect of DNA sequence on the self-assembly with **CPT**, structural details of supramolecular organization are needed. Indeed, besides the electrostatic interactions, which are the main driving forces for the self-assembly of DNA (anionic) and the cationic polymer, other types of intermolecular interactions are at play, such as π-π interactions between the aromatic monomers and the nucleobases (particularly for purine bases), cation-π interactions (polymer cationic side group — nucleobase), anion-π interactions (phosphate — monomer), and possibly H-bonding interactions (monomer — base), which all influence the conformational changes of each chain and the assembly within the complex.

An indirect way to get insights into the polymer-DNA binding modes is to use a competitor molecule which is known to bind at specific sites of DNA. For instance, the molecule 4',6-diamidino-2-phenylindole (**DAPI**) is well-known to bind to the DNA minor-groove, preferentially in AT-base pairs rich regions.[45-48] The interactions between **DAPI** and DNA can also be studied by means of (chir)optical spectroscopy.[49,50] When the achiral **DAPI** molecule fits into the minor-groove of a double-stranded DNA helix, an induced positive CD signal appears in the range where the **DAPI** absorbs (range 300–400 nm), and the UV-Vis spectra show a red-shifted absorption maximum (of around 17 nm) compared to the pure **DAPI**, indicating its planarization upon DNA binding. Starting from complexes dsDNA/**DAPI** (dsDNA being either **dA·dT** or **dR·dR$_{rev}$**, Chart 1), the CD spectra after addition of **CPT** shows that the induced CD signal **DAPI** diminishes in the range 300–400 nm (maximum around 370 nm), see Figure 6. This indicates that the polymer competes with the binding mode of **DAPI**, in other words, the polymer is in close proximity of the minor groove. The intensity of the induced CD signal of **DAPI** is more decreased in the case of **dR·dR$_{rev}$/DAPI/CPT** than for **dA·dT/DAPI/CPT**, which is consistent with the preferential binding of **DAPI** at AT-rich regions.[48,51] The CD spectra in the range where the polymer absorbs still present a ($-$/$+$) bisignate ICD signal for the case of **dR·dR$_{rev}$/DAPI/CPT** and no signal for **dA·dT/DAPI/CPT**, as in the cases without **DAPI**. Altogether, this shows that **CPT** likely surrounds the DNA duplex in a chiral conformation that is only partially binding the minor-groove.

To further decipher the supramolecular structure of dsDNA/**CPT** assemblies in aqueous environment, biomolecular simulations were performed by molecular dynamics (MD) simulations on the following complexes

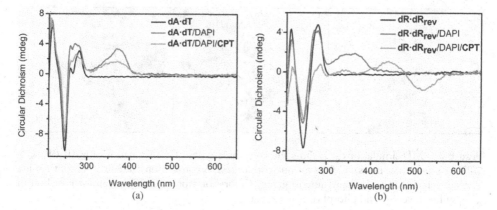

Figure 6: CD spectra of mixtures of DNA duplexes with **DAPI** and **CPT**. (a) **dA·dT** sequence (black line), **dA·dT/DAPI** at 1:4 (red line), and **dA·dT/DAPI/CPT** at 1:4:1 (green line). (b) **dR·dR$_{rev}$** sequence (black line), **dR·dR$_{rev}$/DAPI** at 1:3 (red line), and **dR·dR$_{rev}$/DAPI/CPT** at 1:3:1 (green line). The **DAPI** molar ratio was chosen considering the maximum emission signal in the dsDNA/DAPI titration experiment.

dA·dT/CPT and **dR·dR$_{rev}$/CPT**. The modeled polymer backbone was shorter than in our experiments, i.e. from 10 or 20 monomers-long, for the sake of the computational cost. The MD methodology is shortly described here: the ff99 + bsc0 force field[52] or the ff99 + bsc0 +ε/ζOL1[53] +χOL4[54] force field was used for DNA molecules and the gaff force field [55] was used for **CPT** after a re-parameterization to reproduce bithiophene torsion potential estimated at the MP2/cc-pvdz level. The atomic charges for **CPT** were derived from RESP fits[56] at the HF/6-31G(d) level using the RED Python program, or PyRED, interfaced by R.E.D. Server Development.[57] The system was solvated explicitly using the TIP3P water model.[58]

Whereas an accurate sampling of the DNA/**CPT** supramolecular assembly is tedious due to the conformational diversity of the polymer, in the multiple MD simulations (12 MD runs with production times varying from 50 ns to 300 ns, total of 1200 ns), we observed three distinguishable types of conformations for oligothiophene segments along a polymer chain, as shown in Figure 7 for **dR·dR$_{rev}$/CPT**: (1) the **CPT** backbone in the DNA major groove; (2) the **CPT** backbone in the DNA minor groove; (3) the accommodation of two successive **CPT** monomers (thiophene rings, alkyl chains and phosphonium groups) in the DNA minor-groove, a conformation referred to as "click".

For conformations where the **CPT** backbone is in the DNA major groove, it barely interacts with the DNA bases and a few phosphonium cations located on that segment are close to DNA phosphate groups. In the second set of conformations, the fixation of the entire **CPT** backbone in the DNA

Figure 7:   MD simulations snapshots showing three types of conformations for **dR·dR$_{rev}$ +
CPT** (10 monomer units). Colors for nucleotides (NDB convention): adenine in red, thymine
in blue, cytosine in yellow, guanine in green. Colors for atoms (CPK convention): carbon in
grey, sulfur in yellow and phosphorus in orange.

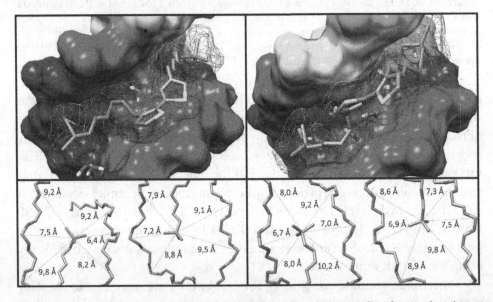

Figure 8:   Top: detailed views of the "click" conformations. Bottom: phosphorus-phosphorus
distances between phosphonium cations and DNA phosphate groups. Colors for nucleotides
(NDB convention): adenine in red, thymine in blue, cytosine in yellow, guanine in green. Colors
for atoms (CPK convention): carbon in grey, sulfur in yellow and phosphorus in orange.

minor-groove is only possible if the **CPT** backbone is in a right-handed
helical conformation. However, this conformation is contradictory to the
experimental CD spectra, which indicate a preferential left-handed polymer
assembly. In the third set of conformations, ("click"), the accommodation
of phosphonium groups in the minor-groove is electrostatically favorable, as
illustrated in Fig. 8. It shows that 5 or 6 DNA phosphate groups are placed
at a distance of less than 11.5 Å of a **CPT** phosphonium group (distances

are measured between phosphorus atoms). Moreover, the van der Waals diameter of tetramethylphosphonium is around 7 Å, which indicates that these cationic groups can fit in the minor groove without important deformation in **dR·dR$_{rev}$** minor groove. However, in MD simulations of **dA·dT/CPT**, this "click" conformation necessitates a larger deformation of the minor-groove, because the **dA·dT** duplex minor-roove is narrower.[59] This is consistent with the differences in induced CD signals of **DAPI** in the above competitor experiments. Multiple "clicks" could form between dsDNA and **CPT**, which could favor particular **CPT** backbone conformations (possibly left-handed) in between the "clicks".

6. Conclusion and Outlook on Structural Aspects of CPE/DNA Assemblies

Cationic conjugated polymers offer many perspectives in biomedical research, noticeably for DNA biosensors for diagnostics, and gene-delivery polyplexes. Together with other reports,[1,9] our studies show that cationic polythiophenes can efficiently be used for DNA-DNA hybridization sensors and SNP detection. However, the detailed understanding of their structure and self-assembly with DNA is key to further optimize their targeted properties, and this involves deciphering the interplay of the interactions between DNA and the cationic polymer. By a careful study on selected oligonucleotide sequences, we have shown that monomer-nucleobase interactions and cation-grooves interactions are important in the (chiral) self-assembly of DNA with cationic polythiophenes. This provides clues for elucidating the effects of non-homogeneity of the assays in DNA-DNA hybridization biosensors involving CPEs. The effects of aggregation should be better understood, possibly by combining coarse-grained models with spectroscopic and scattering techniques.

Acknowledgments

The authors thank their collaborators in Mons and in Montpellier. The authors are grateful to the financial support from the Fonds de la Recherche Scientifique — FNRS under grants CHIRNATES n°1.B333.15F, DRaPo n° 1.A.230.16F and SHERPA n°F.4532.16 as well as computational resources provided by the CECI. The CNRS and the Université de Montpellier are acknowledged for support. M.S. thanks the Pôle Chimie-Fondation Balard for a visiting professor grant within the University of Montpellier.

References

1. H.-A. Ho, A. Najari, M. Leclerc, Optical detection of DNA and proteins with cationic polythiophenes. *Acc. Chem. Res.*, **41(2)**, 168–178 (2008).
2. C. Zhu, L. Liu, Q. Yang, F. Lv, S. Wang, Water-soluble conjugated polymers for imaging, diagnosis, and therapy. *Chem. Rev.*, **112(8)**, 4687–4735 (2012).
3. G. Feng, D. Ding, B. Liu, Fluorescence bioimaging with conjugated polyelectrolytes. *Nanoscale*, **4(20)**, 6150–6165 (2012).
4. G. Feng, J. Liang, B. Liu, Hyperbranched conjugated polyelectrolytes for biological sensing and imaging. *Macromol. Rapid Commun.*, **34(9)**, 705–715 (2013).
5. R. Zhan and B. Liu, Benzothiadiazole-containing conjugated polyelectrolytes for biological sensing and imaging. *Macromol. Chem. Phys.*, **216(2)**, 131–144 (2015).
6. C. Chi, A. Mikhailovsky, G. C. Bazan, Design of cationic conjugated polyelectrolytes for DNA concentration determination. *J. Am. Chem. Soc.*, **129(36)**, 11134–11145 (2007).
7. K. P. R. Nilsson and O. Inganäs, Chip and solution detection of DNA hybridization using a luminescent zwitterionic polythiophene derivative. *Nat. Mater.*, **2(6)**, 419–424 (2003).
8. B. Liu and G. C. Bazan, Methods for strand-specific DNA detection with cationic conjugated polymers suitable for incorporation into DNA chips and microarrays. *Proc. Natl. Acad. Sci. USA*, **102(3)**, 589–593 (2005).
9. B. S. Gaylord, M. R. Massie, S. C. Feinstein, G. C. Bazan, SNP detection using peptide nucleic acid probes and conjugated polymers: Applications in neurodegenerative disease identification. *Proc. Natl. Acad. Sci. USA*, **102(1)**, 34–39 (2005).
10. J. Liang, K. Li, B. Liu, Visual sensing with conjugated polyelectrolytes. *Chem. Sci.*, **4(4)**, 1377–1394 (2013).
11. G. Yang, H. Yuan, C. Zhu, L. Liu, Q. Yang, F. Lv, S. Wang, New conjugated polymers for photoinduced unwinding of DNA supercoiling and gene regulation. *ACS Appl. Mater. Interfaces*, **4(5)**, 2334–7 (2012).
12. X. Feng, F. Lv, L. Liu, Q. Yang, S. Wang, G. C. Bazan, A highly emissive conjugated polyelectrolyte vector for gene delivery and transfection. *Adv. Mater.*, **24(40)**, 5428–5432 (2012).
13. M. L. Davies, P. Douglas, H. D. Burrows, B. Martincigh, M. d. G. a. Miguel, U. Scherf, R. Mallavia, A. Douglas, In depth analysis of the quenching of three fluorene phenylene-based cationic conjugated polyelectrolytes by DNA and DNA bases. *J. Phys. Chem. B*, **118(2)**, 460–469 (2014).
14. Z. Liu, H.-L. Wang, M. Cotlet, DNA sequence-dependent photoluminescence enhancement in a cationic conjugated polyelectrolyte. *Chem. Commun.*, **50(77)**, 11311–11313 (2014).
15. M. Knaapila, T. Costa, V. M. Garamus, M. Kraft, M. Drechsler, U. Scherf, H. D. Burrows, *Macromolecules*, **47**, 4017–4027 (2014).
16. J. Rubio-Magnieto, E. G. Azene, J. Knoops, S. Knippenberg, C. Delcourt, A. Thomas, S. Richeter, A. Mehdi, Ph. Dubois, R. Lazzaroni, D. Beljonne, S. Clément, M. Surin, Self-assembly and hybridization mechanisms of DNA with cationic polythiophene. *Soft Matter*, **11(32)**, 6460–6471 (2015).
17. J. Rubio-Magnieto, A. Thomas, S. Richeter, A. Mehdi, P. Dubois, R. Lazzaroni, S. Clément, M. Surin, Chirality in DNA-pi-conjugated polymer supramolecular structures: insights into the self-assembly. *Chem. Commun.*, **49(48)**, 5483–5485 (2013).

18. F. Xia, X. Zuo, R. Yang, Y. Xiao, D. Kang, A. Vallée-Bélisle, X. Gong, A. J. Heeger, K. W. Plaxco, On the binding of cationic, water-soluble conjugated polymers to DNA: Electrostatic and hydrophobic interactions. *J. Am. Chem. Soc.*, **132(4)**, 1252–1254 (2010).

19. Z. Liu, H.-L. Wang, M. Cotlet, Energy transfer from a cationic conjugated polyelectrolyte to a DNA photonic wire: Toward label-free, sequence-specific DNA sensing. *Chem. Mater.*, **26**, 2900–2906 (2014).

20. I. Charlebois, C. Gravel, N. Arrad, M. Boissinot, M. G. Bergeron, M. Leclerc, Impact of DNA sequence and oligonucleotide length on a polythiophene-based fluorescent DNA biosensor. *Macromol. Biosci.*, **13(6)**, 717–22 (2013).

21. W. Yanyan and L. Bin, Cationic water-soluble polyfluorene homopolymers and copolymers: Synthesis, characterization and their applications in DNA sensing. *Curr. Org. Chem.*, **15(4)**, 446–464 (2011).

22. I. Osaka and R. D. McCullough, Advances in molecular design and synthesis of regioregular polythiophenes. *Acc. Chem. Res.*, **41(9)**, 1202–1214 (2008).

23. P. Sista and C. K. Luscombe, Progress in the synthesis of poly(3-hexylthiophene), Edited by S. Ludwigs, *P3HT Revisited — From Molecular Scale to Solar Cell Devices*, ISBN 978-3-662-45145-8 — Springer Berlin Heidelberg, 1–38 (2014).

24. M. C. Stefan, M. P. Bhatt, P. Sista, H. D. Magurudeniya, Grignard metathesis (GRIM) polymerization for the synthesis of conjugated block copolymers containing regioregular poly(3-hexylthiophene). *Polym. Chem.*, **3(7)**, 1693–1701 (2012).

25. M. Chevrier, J. E. Houston, J. Kesters, N. Van den Brande, A. E. Terry, S. Richeter, A. Mehdi, O. Coulembier, Ph. Dubois, R. Lazzaroni, B. Van Mele, W. Maes, R. C. Evans, S. Clément, Self-assembled conjugated polyelectrolyte-surfactant complexes as efficient cathode interlayer materials for bulk heterojunction organic solar cells. *J. Mater. Chem. A*, **3(47)**, 23905–23916 (2015).

26. A. Thomas, J. E. Houston, N. Van den Brande, J. De Winter, M. Chevrier, R. K. Heenan, A. E. Terry, S. Richeter, A. Mehdi, B. Van Mele, Ph. Dubois, R. Lazzaroni, P. Gerbaux, R. C. Evans, S. Clément, All-conjugated cationic copolythiophene "rod-rod" block copolyelectrolytes: synthesis, optical properties and solvent-dependent assembly. *Polym. Chem.*, **5(10)**, 3352–3362 (2014).

27. S. Clément, A. Tizit, S. Desbief, A. Mehdi, J. De Winter, P. Gerbaux, R. Lazzaroni, B. Boury, Synthesis and characterisation of π-conjugated polymer/silica hybrids containing regioregular ionic polythiophenes. *J. Mater. Chem.*, **21(8)**, 2733–2739 (2011).

28. C. Chi, A. Chworos, J. Zhang, A. Mikhailovsky, G. C. Bazan, Anatomy and growth characteristics of conjugated polyelectrolyte/DNA aggregates. *Adv. Funct. Mater.*, **18(22)**, 3606–3612 (2008).

29. H.-A. Ho, M. Boissinot, M. G. Bergeron, G. Corbeil, K. Doré, D. Boudreau, M. Leclerc, Colorimetric and fluorometric detection of nucleic acids using cationic polythiophene derivatives. *Angew. Chem. Int. Ed.*, **41(9)**, 1548–1551 (2002).

30. H.-A. Ho and M. Leclerc, Optical sensors based on hybrid aptamer/conjugated polymer complexes. *J. Am. Chem. Soc.*, **126(5)**, 1384–1387 (2004).

31. H. A. Ho, K. Doré, M. Boissinot, M. G. Bergeron, R. M. Tanguay, D. Boudreau, M. Leclerc, Direct molecular detection of nucleic acids by fluorescence signal amplification. *J. Am. Chem. Soc.*, **127(36)**, 12673–12676 (2005).

32. M. Béra Abérem, A. Najari, H. A. Ho, J. F. Gravel, P. Nobert, D. Boudreau, M. Leclerc, Protein detecting arrays based on cationic polythiophene–DNA-aptamer complexes. *Adv. Mater.*, **18(20)**, 2703–2707 (2006).

33. J. De Winter, G. Deshayes, F. Boon, O. Coulembier, Ph. Dubois, P. Gerbaux, MALDI-ToF analysis of polythiophene: Use of trans-2-[3-(4-t-butyl-phenyl)-2-methyl-2-propenylidene]malononitrile — DCTB — as matrix. *J. Mass Spectrom.*, **46(3)**, 237–246 (2011).

34. K. P. Nilsson, J. Rydberg, L. Baltzer, O. Inganäs, Twisting macromolecular chains: Self-assembly of a chiral supermolecule from nonchiral polythiophene polyanions and random-coil synthetic peptides. *Proc. Natl. Acad. Sci. USA*, **101(31)**, 11197–202 (2004).

35. M. Surin, P. G. A. Janssen, R. Lazzaroni, Ph. Leclère, E. W. Meijer, A. P. H. J. Schenning, Supramolecular organization of ssDNA-Templated π-conjugated oligomers via hydrogen bonding. *Adv. Mater.*, **21**, 1126–1130 (2009).

36. A. Digennaro, H. Wennemers, G. Joshi, S. Schmid, E. Mena-Osteritz, P. Bauerle, Chiral suprastructures of asymmetric oligothiophene-hybrids induced by a single proline. *Chem. Commun.*, **49(93)**, 10929–10931 (2013).

37. N. Berova, P. L. Polavarapu, K. Nakanishi, R. W. Woody, eds. *Comprehensive Chiroptical Spectroscopy. Applications in Stereochemical Analysis of Synthetic Compounds, Natural Products, and Biomolecules*. ed. Wiley (2012).

38. H.-A. Ho, M. Béra-Abérem, M. Leclerc, Optical sensors based on hybrid DNA/conjugated polymer complexes. *Chem. Eur. J.*, **11(6)**, 1718–1724 (2005).

39. L. Yang, M. Zhao, R. Zhang, J. Dong, T. Zhang, X. Zhan, G. Wang, Synthesis and fluorescence study of a quaternized copolymer containing pyrene for DNA-hybridization detection. *ChemPhysChem.*, **13(18)**, 4099–4104 (2012).

40. M. L. Davies, P. Douglas, H. D. Burrows, B. Martincigh, M. da Graca Miguel, U. Scherf, R. Malavia, A. Douglas, In depth analysis of the quenching of three fluorene phenylene-based cationic conjugated polyelectrolytes by DNA and DNA bases. *J. Phys. Chem. B*, **118**, 460 (2014).

41. B. S. Shastry, SNP alleles in human disease and evolution. *J. Hum. Genet.*, **47(11)**, 561–566 (2002).

42. C. Whibley, P. D. P. Pharoah, M. Hollstein, p53 polymorphisms: Cancer implications. *Nat. Rev. Cancer*, **9(2)**, 95–107 (2009).

43. N. Rivlin, R. Brosh, M. Oren, V. Rotter, Mutations in the p53 tumor suppressor gene: Important milestones at the various steps of tumorigenesis. *Genes Cancer*, **2(4)**, 466–474 (2011).

44. D. Gerion, F. Chen, B. Kannan, A. Fu, W. J. Parak, D. J. Chen, A. Majumdar, A. P. Alivisatos, Room-temperature single-nucleotide polymorphism and multiallele DNA detection using fluorescent nanocrystals and microarrays. *Anal. Chem.*, **75(18)**, 4766–4772 (2003).

45. T. A. Larsen, D. S. Goodsell, D. Cascio, K. Grzeskowiak, R. E. Dickerson, The structure of DAPI bound to DNA. *J. Biomol. Struct. Dyn.*, **7(3)**, 477–491 (1989).

46. T. Härd, P. Fan, D. R. Kearns, A fluorescence study of the binding of Hoechst 33258 and DAPI to halogenated DNAs. *Photochem. Photobiol.*, **51(1)**, 77–86 (1990).

47. W. D. Wilson, F. A. Tanious, H. J. Barton, R. L. Jones, K. Fox, R. L. Wydra, L. Strekowski, DNA sequence dependent binding modes of 4′,6-diamidino-2-phenylindole (DAPI). *Biochemistry*, **29(36)**, 8452–8461 (1990).

48. M. J. Hannon, Supramolecular DNA recognition. *Chem. Soc. Rev.*, **36(2)**, 280–295 (2007).

49. H.-K. Kim, J.-M. Kim, S. K. Kim, A. Rodger, B. Nordén, Interactions of intercalative and minor groove binding ligands with triplex poly(dA)·[poly(dT)]2 and with duplex poly(dA)·poly(dT) and poly[d(A-T)]2 studied by CD, LD, and normal absorption. *Biochemistry*, **35(4)**, 1187–1194 (1996).

50. B. Jin, H. M. Lee, Y.-A. Lee, J. H. Ko, C. Kim, S. K. Kim, Simultaneous binding of meso-tetrakis(N-methylpyridinium-4-yl)porphyrin and 4′,6-diamidino-2-phenylindole at the minor grooves of poly(dA)·poly(dT) and poly[d(A−T)$_2$]: Fluorescence resonance energy transfer between DNA bound drugs. *J. Am. Chem. Soc.*, **127(8)**, 2417–2424 (2005).

51. J. Kapuściński and W. Szer, Interactions of 4′, 6-diamidine-2-phenylindole with synthetic polynucleotides. *Nucleic Acids Res.*, **6(11)**, 3519–3534 (1979).

52. A. Pérez, I. Marchán, D. Svozil, J. Sponer, T. E. Cheatham, C. A. Laughton, M. Orozco, Refinement of the AMBER force field for nucleic acids: Improving the description of α/γ conformers. *Biophys. J.*, **92(11)**, 3817–3829 (2007).

53. M. Zgarbová, F. J. Luque, J. Šponer, T. E. Cheatham, M. Otyepka, P. Jurečka, Toward improved description of DNA backbone: Revisiting epsilon and zeta torsion force field parameters. *J. Chem. Theory Comput.*, **9(5)**, 2339–2354 (2013).

54. M. Krepl, M. Zgarbová, P. Stadlbauer, M. Otyepka, P. Banáš, J. Koča, T. E. Cheatham, P. Jurečka, J. Šponer, Reference simulations of noncanonical nucleic acids with different χ variants of the AMBER force field: Quadruplex DNA, quadruplex RNA, and Z-DNA. *J. Chem. Theory Comput.*, **8(7)**, 2506–2520 (2012).

55. J. Wang, R. Wolf, J. Caldwell, P. Kollman, D. Case, Development and testing of a general amber force field. *J. Comput. Chem.*, **25(9)**, 1157–1174 (2004).

56. C. I. Bayly, P. Cieplak, W. Cornell, P. A. Kollman, A well-behaved electrostatic potential based method using charge restraints for deriving atomic charges: The RESP model. *J. Phys. Chem.*, **97(40)**, 10269–10280 (1993).

57. E. Vanquelef, S. Simon, G. Marquant, E. Garcia, G. Klimerak, J. Delepine, P. Cieplak, F.-Y. Dupradeau, R.E.D. Server: A web service for deriving RESP and ESP charges and building force field libraries for new molecules and molecular fragments. *Nucleic Acids Res.*, **39(suppl. 2)**, W511–W517 (2011).

58. W. L. Jorgensen, J. Chandrasekhar, J. D. Madura, R. W. Impey, M. L. Klein, Comparison of simple potential functions for simulating liquid water. *J. Chem. Phys.*, **79(2)**, 926–935 (1983).

59. C. Yoon, G. G. Privé, D. S. Goodsell, R. E. Dickerson, Structure of an alternating-B DNA helix and its relationship to A-tract DNA. *Proc. Natl. Acad. Sci. USA*, **85(17)**, 6332–6336 (1988).

Chapter 6

High Throughput *in situ* Scattering of Roll-To-Roll Coated Functional Polymer Films

Jens Wenzel Andreasen

Department of Energy Conversion and Storage
Technical University of Denmark
Frederiksborgvej 399
4000 Roskilde, Denmark
jewa@dtu.dk

The development of conjugated polymers for organic electronics and photovoltaics has relied heavily on advanced X-ray scattering techniques almost since the earliest studies in the field. Almost from the beginning, structural studies focused on how the polymers self-organize in thin films, and the relation between chemical configuration of the polymer, structure and performance. This chapter presents the latest developments where structural analysis is applied as *in situ* characterization of structure formation during roll-to-roll coating of photoactive layers for solar cells.

1. X-ray Scattering Techniques for Large Scale Characterization of Nano-Structures in Thin Films

X-ray characterization has played an important role throughout the development of conductive polymers, beginning with the structural study of bulk polyacetylene, establishing its very high crystallinity and drawing-induced texture.[1] With the advent of soluble polythiophene derivatives, structural studies of solution-cast, self-supporting films[2,3] paved the way for establishing correlations between electronic performance and crystalline texture in thin films of poly(3-hexylthiophene).[4] The latter study showed that regioregularity, molecular weight and casting method may all have significant impact on molecular self-assembly in thin films in terms of resulting crystallinity and texture.

In contrast to the direct space techniques, most X-ray techniques based on scattering are bulk averaging probes, sampling a representation of the nanostructure in reciprocal space, essentially the Fourier transform of the electron density distribution in the sample. The coherent scattering from different points in the sample ads up according to amplitude and relative phase. The phase difference of the coherently scattered radiation depends on the relative location of the scattering centers. In the actual experiment, scattered intensities are detected, i.e. the absolute square of the scattering amplitude, and phase information is lost. As a consequence there is in general no model-free way of retrieving the structure from the scattering data. In many cases, several different structural models will in fact produce the same scattering pattern, i.e. there is no unique solution. The advantage of the techniques is that they provide statistically significant data for a large volume of the sample, e.g. the entire thickness of a film and an area of up to several millimetres squared. With appropriate assumptions based on *a priori* information about the sample, meaningful quantitative characteristics of the sample may often be derived, suitable for correlation with performance data for a macroscopic device. Another advantage of bulk averaging scattering techniques is that they are particularly well suited for fast characterization of large volumes of materials, thus lending themselves to time-resolved *in situ* applications. Some of the most commonly applied techniques are outlined in the following.

Small Angle X-ray Scattering (SAXS) is used for characterizing structures ranging in sizes from a few Ångström to close to a micron, overlapping with light scattering techniques. It is typically applied in transmission, but may also be used for thin film characterization by selecting a suitably shallow X-ray incidence angle with respect to the surface, below the critical angle for total reflection from the substrate, so that the beam is only scattered from the film. This variation of the technique is usually referred to as Grazing Incidence SAXS, or GISAXS and is a very efficient application of SAXS to thin films because of the large sample volume interacting with the beam (due to the long beam footprint at shallow angles) and the suppression of scattering from the substrate (Fig. 1). The data analysis is somewhat more challenging than for standard transmission SAXS, especially for scattering vectors in the plane of the surface normal, because of interference between the scattered radiation from the thin film and the radiation reflected from the substrate and film surface.

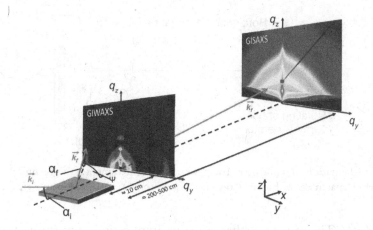

Figure 1:   The scattering geometry that applies to GISAXS and GIWAXS measurements. The incident beam wave vector, $\vec{k_i}$ impinges at a shallow angle α_i with respect to the substrate surface. The scattered wave vector $\vec{k_f}$ may be recorded at a short distance (5–20 cm) for large exit angles in the surface normal plane, α_i and ψ_f in the substrate plane (GIWAXS). At longer detection distances of 2 to 10 meters, smaller exit angles are recorded, corresponding to the GISAXS signal.[5] (Reproduced with permission of the International Union of Crystallography.)

Wide Angle X-ray Scattering (WAXS) covers techniques that probe larger scattering vectors, typically at scattering angles from 5°, thus sensitive to atomic order as observed in crystals. Similar to SAXS it may be applied at a grazing incidence (GIWAXS) and is as such developed from surface diffraction techniques, first applied at synchrotron radiation sources. The first applications of the technique were mainly in the field of inorganic crystal surfaces, their reconstruction and the structure of epitaxial overlayers,[6] but using 2D detectors and specially designed low-background instrumentation like the GIWAXS setup at DTU Energy[7] sketched in Fig. 2, the technique has found wide-spread application in the field of organic electronics and photovoltaics for characterization of crystallinity and texture.[4,8–14]

X-ray reflectometry is a powerful technique for determining density variations along the surface normal of thin films down to sub-nm scale. To obtain the highest detail, very smooth and planar films are required, something which is often not realized when coating polymeric films.[15] The original implementation of reflectometry relies on a highly collimated beam and a scanning point detector to record the specularly reflected X-ray beam as function of incidence angle. To allow faster acquisition for *in situ* applications, alternative approaches include energy dispersive reflectometry,[16] use of area detectors[17] and convergent beams coupled with an area detector to enable acquisition of the entire reflectivity curve in one exposure for a fixed sample

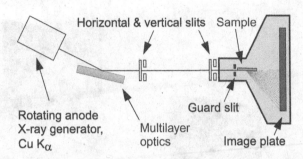

Figure 2:   Schematic layout of the low background GIWAXS camera at the Technical University of Denmark, Department of Energy Conversion and Storage. (Adapted from Ref. 7.)

orientation.[18] This approach has obvious implications for the potential for integration of reflectometry with roll-to-roll processing, not least because it is anticipated that the small footprint obtained by coupling the high-convergence optics with a microfocus source, will reduce the negative effects of roughness or thickness inhomogeneities on the macroscopic level, substantially.

In combination, these X-ray scattering techniques can provide comprehensive details of thin film structures from the atomic level to micron-scale: Materials that crystallize can be characterized in terms of their crystal structure, crystallinity, crystalline microstructure (strain and defects) and texture (crystallite size, habit and preferred orientation) using GIWAXS.[19] Reflectometry and GISAXS are sensitive to variations in electron density on scales from a few nanometres up to microns, thus complementing GIWAXS excellently with the ability to characterize nanostructure, even in completely amorphous materials.

It is by virtue of these abilities, that X-ray scattering techniques lend themselves particularly well to structural studies of conjugated polymers that cover the entire range of conformations from completely amorphous to highly crystalline, with various degrees of semicrystallinity in between. With the high penetration power of X-rays and relative ease of application in ambient environment, it naturally follows to consider the extension to *in situ* studies of structure formation during processing with methods relevant to mass production.

2.   *In situ* Studies of Structural Formation

Like most research in functional conjugated thin films, *in situ* X-ray characterization of the coating process have developed from studies of drop and spin-casting on glass that represents small-scale laboratory systems, relatively easy to implement as a very controlled environment for detailed study of structure formation and development.[20,21] To really benefit from the processability of polymers, however, we need to understand the dynamics of fast, scalable processing methods such as roll-to-roll (R2R) coating on flexible substrates. A fast coating method like slot-die coating, for instance, certainly constitutes a very different system with respect to spin coating, in terms of drying time and substrate properties.[22]

One of the earliest implementations of *in situ* X-ray scattering for the study of a scalable coating technique applied to a conjugated polymer system was reported by Sanyal and Schmidt-Hansberg.[23–26] They designed a doctor blade coating system that can coat a short length of film on a solid substrate and allow monitoring the film thickness and X-ray scattering while the film dries at a synchrotron GIWAXS beam line (Fig. 3).

With this setup they were able to evaluate the effect of substrate temperature on the drying and crystallization time and the relation to final crystallinity and texture of the conjugated polymer poly(3-hexylthiophene) in a blend with [6,6]-phenyl-C61 butyric acid methyl ester (P3HT:PCBM), relevant as active layer for polymer solar cells.[23] They also established the relation between blend ratio and P3HT crystallization kinetics and the resulting texture.[24] The interaction between P3HT and PCBM during drying was further investigated and modified by using a mixed solvent where the minority component is a poor solvent for both components, causing the formation of aggregates in the still wet film.[25] For the important class of low bandgap polymers, it was possible

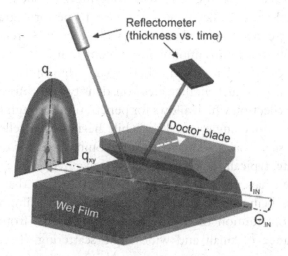

Figure 3: Schematic of the experimental set-up for real-time grazing incidence X-ray scattering and laser reflectometry. (Reproduced from Ref. 23 with permission from John Wiley and Sons, Copyright © 2011 WILEY-VCH Verlag GmbH & Co. KGaA, Weinheim.)

to show with blends of poly{[4,40-bis(2-ethylhexyl)dithieno(3,2-b;20,30-d)silole]-2,6-diyl-alt-(2,1,3-benzothidiazole)-4,7-diyl} (PSBTBT) and [6,6]-phenyl C_{71}-butyric acid methyl ester ($PC_{71}BM$) that crystallization in this system is preferentially happening in the bulk solution and to a much lesser degree at the film interfaces as observed for P3HT:PCBM.[26] This makes the PSBTBT:PC_{71}BM system much less sensitive to the effect of substrate temperature, thus allowing faster drying, suitable for fast R2R coating. More recently, another implementation of a coating method that is akin to real scalable roll-to-roll coating was demonstrated by Pröller *et al.*,[27] but also applied to short lengths of solid substrates (silicon wafer), thus still somewhat removed from the conditions of fast processing.

These studies were clearly important strides towards a better understanding of the drying kinetics in coating of polymer solar cell active layers, but could just as well have been carried out by drop casting, as these setups only allowed what corresponds to small batch experiments on solid substrates. In order to allow investigation of structure formation during real R2R coating, it is necessary to be able to record X-ray scattering from films coated on flexible substrates.

3.   High Throughput Roll-To-Roll X-ray Characterization

As already mentioned, experimenters have preferentially used float glass or silicon wafers as substrates for X-ray studies of conjugated polymer thin films. This is because of the very low roughness of such substrates on a macroscopic scale that makes it relatively easy to perform grazing incidence X-ray scattering experiments where the incident beam is reflected from the surface of the substrate, thus minimizing its contribution to the total scattered signal. Compared to polymer and other organic thin films, a glass or silicon substrate will have a much higher electron density and thus a higher critical angle for total reflection which allows for penetration through the organic film at a shallow incidence angle ($\sim 0.15°$) while having total reflection from the substrate. The situation is somewhat more complicated for depositions on a flexible substrate, typically a polymer such as polyethylene terephthalate (PET) which will often have an electron density very close to that of the organic material that is being coated. In this case, it is nearly impossible to avoid a substantial contribution to the total scattering signal from the substrate, both in the range of small and wide angle scattering. There are however also considerable advantages by working with coatings on a flexible substrate. Besides studying the coated thin films under the actual processing conditions that allow scaling to mass production, it also facilitates large scale analysis of processing parameters with high resolution. Böttiger *et al.*[28] demonstrated this as a structural analysis extension to an earlier large scale solar cell device analysis experiment, where constituent concentration ratio and film thickness was varied over tens of meters of roll-coated devices.[29] In this manner, "recording" thousands of experiment variations along the substrate foil, high-resolution analysis of parameter space is allowed by "playing back" the coating, either on a fast testing device to record solar cell device parameters (IV-curves), or through a high-flux X-ray beam to record the small- and wide-angle scattering from the coated film (Fig. 4). Depending on X-ray flux, acquisition times down to a few hundred milliseconds can be realized corresponding to a spatial resolution along the coated foil of a millimeter to a centimetre for foil speeds between 0.5 to 5 m/min. For an experiment where a component concentration is varied from 0% to 100%, this corresponds to a resolution in concentration parameter space of better than 0.01%, or better than 0.001% for the case of additive concentration, varied from 0% to 5%.[28] This is well below what can be expected in mixing accuracy from stepper-controlled syringe pumps and uncertainties in the flow dynamics of the microchannel solution mixing structure. Accepting a lower resolution of parameter space, i.e. fewer measurements along the coated foil, for instance at every 5 cm, similar experiments can be completed in a few days with an optimized laboratory instrument[30] where acceptable counting statistics can be achieved in 1000 s exposures.

With this experimental design, it was possible to investigate the effect of variations of several important processing parameters for coating on flexible substrate, with a resolution hitherto unheard of. Specifically, the effect of three important parameters was probed in great detail:

1. Active layer thickness. By varying the concentration of a P3HT:PCBM 1:1 stock solution with respect to pure chlorobenzene, the dry thickness of the active layer could be varied from 140 to 400 nm. It was found that highly textured (edge-on) P3HT crystallites dominate in the thinnest films, below 270 nm thickness, whereas increasing amounts of untextured crystallites of P3HT are found in thicker films, adopting a slightly different structural form with a shorter lamellar spacing, indicating that crystallites are forming in solution, at higher concentration levels.

2. Donor/acceptor ratio. Through variation of the P3HT to PCBM ratio, it was found that P3HT does not crystallize when slot-die coated on flexible substrate at ratios below 1:2 with respect to PCBM. At almost exactly 1:1 ratio, the P3HT crystallites show a pronounced minimum in crystallite size.

3. Solvent additive. Addition of a high boiling point solvent (chloronaphtalene) affects the P3HT crystallization significantly, with a clear optimum for highest crystallinity and largest crystals at 2.0 vol% addition.

Figure 4:   The R2R unit in action at the cSAXS beam line at the Swiss Light Source, Paul Scherrer Institut. The unit is mounted on the sample stage hexapod to allow accurate positioning and control of the incidence angle. The beam flight tube exit is seen on the right, and the SAXS flight tube entry on the left. On the substrate foil, the red active layer film can be seen as well as a couple of bar codes used to keep track of the experiment sample ID, as encoded during coating. The scanning pattern of the bar code reader laser can just be made out over the bar code.[28] (Reproduced by permission of The Royal Society of Chemistry.)

When such structural information is combined with highly resolved measurements of the solar cell device performance, we may draw important conclusions on fundamental parameters governing relations between nanostructure and performance. We note for instance that the observed minimal P3HT crystallite size at 1:1 mixing ratio with PCBM coincides with the optimal power conversion efficiency found in several other studies of this materials system.

With this high-throughput study of solar cell active layers, coated on flexible substrate, the road was paved for the logical next step:

4.   *In situ* X-ray Characterization of Polymer Solar Cell Active Layer Drying

As mentioned above, the technique of grazing incidence X-ray scattering from thin film coatings on flexible substrates, can also be implemented at an optimized laboratory scale instrument, as the one described by Kehres *et al.*[30] Here, a flux of 10^7–10^8 photons in a 1 mm beam can be achieved, and although this is 4 to 5 orders of magnitude below what is routinely achieved at 3rd generation synchrotron sources, it can be sufficient to realize *in situ* type

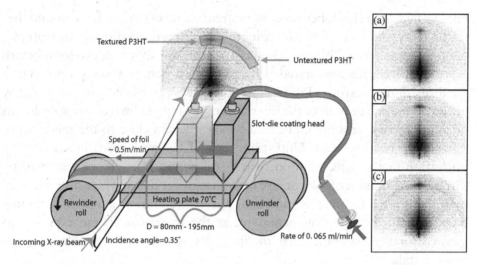

Figure 5: Schematic of the micro-roll coater that can be integrated with either laboratory or synchrotron based SAXS or WAXS instruments. To the right are shown 3 examples of GISAXS measurements integrated from 20 to 30 meters of coating at distances from the coating head of (a) 120 mm, (b) 195 mm and (c) re-rolled in dry condition, corresponding to 14 s, 23 s and 1 h drying time, respectively. The X-ray beam originates at a rotating anode and enters an evacuated flight tube after hitting the film before encountering the detector 1 m further downstream. Using a syringe the P3HT/PCBM solution is pumped through the coating head with a constant rate, while the rewinder pulls the foil from the unwinder, thereby depositing a film of the mixture on the foil. The solution supplied to the coater head may be mixed in a microchannel mixing structure, fed by dual syringes to create variations in constituent concentrations during measurements (Rossander, Zawacka, Dam, Krebs & Andreasen, AIP Advances, 4, 087105 (2014); used in accordance with the Creative Commons Attribution (CC BY) license).

experiments, by careful experimental design and because the more divergent beam allows more crystallites to be in diffraction condition at a fixed incidence angle (at the cost of a lower resolution in reciprocal space). By measuring at a constant position with respect to a slot-die coating head, while continuously coating at about 0.5 m/min, taking X-ray exposures for as long as the substrate foil is running, it is possible to obtain GISAXS/GIWAXS data with good statistics by averaging over data from 40 to 60 minutes, corresponding to 20 to 30 meters of coating. By adjusting the distance of the slot-die coating head with respect to the X-ray sampling position, various drying stages of the coated film can be structurally characterized in this manner, from the wet film just next to the coating head, to drier states further away. More details of drying kinetics can be probed by adjusting the heating plate temperature below the substrate and the speed of the foil (Fig. 5).

With experiments designed like this, it was possible to study the formation of crystalline structure during drying of thin films coated from 1:1 blends of

P3HT/PCBM in chlorobenzene, at different stages of drying. It was found that textured crystallites are formed earlier than 9 s after deposition of the material, but untextured crystallites takes longer to form, and appear not to form before nearly all solvent has evaporated.[31] Based on these results, it was proposed that the formation of textured P3HT crystallites takes place almost immediately by nucleation at one or both of the interfaces of the liquid film (substrate/solution or solution/air), possibly influenced by shear forces close to the meniscus of the slot-die coating head. Untextured crystallites are formed much later by nucleation in the bulk solution, when concentration is elevated sufficiently through evaporation of solvent. We may thus expect that layers of different final morphology are formed, as caused by the concentration gradient stemming from P3HT crystallization beginning at the interfaces. These experiments showed for the first time, that *in situ* X-ray scattering is possible, even on moving foils.

Earlier studies of the effect of addition of a high boiling point solvent, chloronaphthalene[28] were extended to *in situ* investigations following the same principle and confirmed the very early formation of textured P3HT crystallites at drying times shorter than 6.5 s[32] with 65% of the final amount of textured crystallites formed already at this stage. By application of the GISAXS experiment *in situ* it was possible to show that addition of 3% chloronaphthalene to the solution allowed the formation of untextured crystallites to complete in the first drying stage, without further thermal annealing.

5.   Aiding Scalable Polymer Solar Cell Production by *in situ* X-ray Characterization

With the developments described above, X-ray scattering applied *in* or *ex situ* has become a tool that may aid in actual large scale solution-based production of devices, similar to the introduction of X-ray metrology as known from the solids-based inorganic semiconductor industry. The processing of multilayer devices from solution is inherently complex (Fig. 6) and requires judicial choice of solvents to avoid redissolving already coated layers. For multijunction solar cells the number of functional layers quickly adds up, e.g. to 14 layers for a tandem polymer solar cell. To aid in the design and manufacture of such a device, *in situ* X-ray scattering has been applied to determine the optimal combination of materials and solvents constituting the intermediate recombination layer between the two active layers.[33] By using the diffraction peak corresponding to the lamellar stack of P3HT as a structural marker, it was possible to verify that all layers of the intermediate layer, consisting

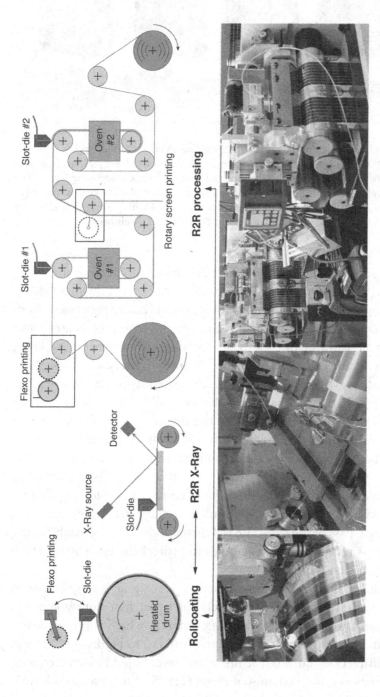

Figure 6: Single roll (top left) and inline roll-to-roll processing (top right) shown schematically (above) with corresponding photographs during coating operation. The junction and process development by sequential stacking of all layers without making surface contact using a roll coater (left) along with the small roll-to-roll X-ray machine (middle) where coating was carried out while the wet, drying and dry film could be probed using Grazing Incidence Small Angle X-ray Scattering (GISAXS). Inline coating using two fully automated slot-die coating stations and two ovens are shown (right) while the intermediate layer comprising PEDOT:PSS and ZnO was prepared. Reproduced from Ref. 33 with permission from the Royal Society of Chemistry.

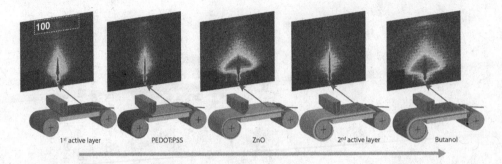

Figure 7:   Small angle X-ray scattering data, corresponding to coating of (from left to right): 1st active layer, PEDOT:PSS and ZnO intermediate layers, 2nd active layer and finally, a layer of butanol, testing the effect of a wet film that is not a solvent for the underlying layers. The position of the 100 reflection from P3HT corresponding to the lamellar stacking is marked in the scattering pattern of the 1st active layer. Adapted from Ref. 34 with permission from John Wiley and Sons, © 2014 WILEY-VCH Verlag GmbH & Co. KGaA, Weinheim.

of PFN (Poly[(9,9-bis(3'-((N,N-dimethyl)-N-ethylammonium)propyl)-2,7-fluorene)-alt-2,7-(9,9-dioctylfluorene)] Dibromide), PEDOT:PSS (poly(3,4-ethylenedioxythiophene) polystyrene sulfonate) and ZnO, processed from respectively methanol, water/isopropanol and acetone could be coated onto the first active layer without disturbing it. Coating the second active layer, however, from an organic solvent like xylene, chlorobenzene, dichlorobenzene or chloroform leads to destruction of the already coated layer. The *in situ* X-ray scattering experiment showed that the polymer is dissolved during coating of the second active layer, as evidenced by the disappearance of the lamellar diffraction peak. This is not evident from an *ex situ* characterization by itself, because the first active layer polymer recrystallizes again. It turns out that the key to preserving the structural integrity of the intermediate layer stack, is to minimize mechanical disturbances in the form of contact as the foil is rolled up in the case of discrete layer coating. With inline coating, where several layers are coated in a single pass of the substrate foil through the coating machine, the intermediate layer is rendered robust enough to protect the underlying active layer from solvent penetration.

To rule out the possibility that the diffraction peak signal is simply masked by scattering and absorption of the relatively thick wet film during coating, the same experiment was carried out coating butanol, which is not a solvent for the first active layer, on top of the intermediate layer (Figure 7).[34] It was shown that the diffraction signal of the first active layer could be detected with just slight perturbation during coating of PEDOT:PSS from aqueous solution, also while coating ZnO from acetone, but disappeared due to solvation during coating of the second active layer from chlorobenzene. The strong small angle scattering from ZnO nanoparticles was also significantly reduced at this stage indicating that the porous structure is complete solvated, thus reducing scattering contrast. After drying and recrystallization, the P3HT diffraction signal was again clearly distinguishable, even while coating butanol on top of the entire stack.

6.   Conclusions and Outlook

The application of scalable processing methods like roll-to-roll coating for devices based on conjugated polymers is being adopted with increasing intensity by research groups around the world, especially in the field of polymer solar cells. Recently, *in situ* X-ray scattering has followed step, allowing high-throughput analysis of the effect of fine-grained variation of processing parameters and inline analysis of the kinetics of structure formation. These methods are rapidly spreading and gaining acceptance in the research community where the speed of roll-to-roll manufacturing is accentuating the need for fast analysis techniques. Most recently, we have seen an implementation similar to the one described above, where a temporal resolution of 40 ms was realized and applied to a drying study of P3HT/PCBM[35] and for comparison with a low bandgap copolymer of 4,5-bis(2-hexyldecyloxy)benzo[2,1-b:3,4-b′]dithiophene (BDT) and dithienylthiazolo[5,4-d]thiazole (TTz) units, (PBDTTTz-4) in blends with PCBM (Ref. 36).

Although many studies of polymer bulk heterojunctions report X-ray scattering data acquired in the small angle range, analyses of non-crystalline, nanostructure from GISAXS are relatively few. Some examples includes the study of blends of MEH-PPV with P3HT where Müller-Buschbaum and co-workers used a combination of GISAXS and X-ray reflectometry for a comprehensive study of optimal blend ratios and film thickness and the relation to nanostructure of the blend[37] and the study of photophysics and the effects of annealing on the nanostructure of P3HT/F8TBT blends by McNeill and co-workers.[38] But, besides simple observations like significant changes in overall scattering strength, signifying the onset of phase separation,[28] or the above mentioned effect of ZnO nanoparticle solvation, analyses of nanostructure in thin films coated on flexible substrates are still lacking. The analysis is complicated by the overlapping contributions of scattering from the substrate and other nanostructured layers that may dominate completely, but this is nevertheless a subject of study that may have considerable impact on the development of electronic devices based on conjugated polymers. Till now, much attention has been devoted to the (semi)crystalline structure of polymers, mainly because it is the most accessible in scattering experiments, but little is still known about the role of nanostructure and how it may be controlled and optimized in an R2R scalable setting. This will likely be an area where *in situ* GISAXS and perhaps even *in situ* reflectometry will provide important new information.

References

1. T. Akaishi, K. Miyasaka, K. Ishikawa, H. Shirakawa, S. Ikeda, Crystallinity of bulk polyacetylene. *J. Polym. Sci. Polym. Phys. Ed.*, **18(4)**, 745–750 (1980).
2. M. J. Winokur, D. Spiegel, Y. Kim, S. Hotta, A. J. Heeger, Structural and absorption studies of the thermochromic transition in poly(3-hexylthiophene). *Synth. Met.*, **28(1–2)**, C419–C426 (1989).
3. T. J. Prosa, M. J. Winokur, J. Moulton, P. Smith, A. J. Heeger, X-ray-diffraction studies of the three-dimensional structure within iodine-intercalated poly(3-octylthiophene). *Phys. Rev. B*, **51(1)**, 159–168 (1995).
4. H. Sirringhaus, P. J. Brown, R. H. Friend, M. M. Nielsen, K. Bechgaard, B. M. W. Langeveld-Voss, A. J. H. Spiering, R. A. J. Janssen, E. W. Meijer, P. Herwig, D. M. de Leeuw, Two-dimensional charge transport in self-organized, high-mobility conjugated polymers. *Nature*, **401(6754)**, 685–688 (1999).
5. A. Hexemer, P. Müller-Buschbaum, Advanced grazing-incidence techniques for modern soft-matter materials analysis. *IUCrJ*, **2(Pt 1)**, 106–25 (2015).
6. R. Feidenhans'l, Surface-structure determination by X-ray-diffraction. *Surf. Sci. Rep.*, **10(3)**, 105–188 (1989).
7. D. Apitz, R. P. Bertram, N. Benter, W. Hieringer, J. W. Andreasen, M. M. Nielsen, P. M. Johansen, K. Buse, Investigation of chromophore-chromophore interaction by electro-optic measurements, linear dichroism, X-ray scattering, density-functional calculations. *Phys. Rev. E*, **72(3)**, 036610, 10 p. (2005).
8. J. W. Andreasen, M. Jorgensen, F. C. Krebs, A route to stable nanostructures in conjugated polymers. *Macromolecules*, **40(22)**, 7758–7762 (2007).
9. E. Pouzet, V. De Cupere, C. Heintz, J. W. Andreasen, D. W. Breiby, M. M. Nielsen, P. Viville, R. Lazzaroni, G. Gbabode, Y. H. Geerts, Homeotropic alignment of a discotic liquid crystal induced by a sacrificial layer. *J. Phys. Chem. C*, **113(32)**, 14398–14406 (2009).
10. H. N. Tsao, D. Cho, J. W. Andreasen, A. Rouhanipour, D. W. Breiby, W. Pisula, K. Müllen, The influence of morphology on high-performance polymer field-effect transistors. *Adv. Mater.*, **21(2)**, 209–212 (2009).
11. C. M. Duffy, J. W. Andreasen, D. W. Breiby, M. M. Nielsen, M. Ando, T. Minakata, H. Sirringhaus, High-mobility aligned pentacene films grown by zone-casting. *Chem. Mater.*, **20(23)**, 7252–7259 (2008).
12. F. C. Krebs, Y. Thomann, R. Thomann, J. W. Andreasen, A simple nanostructured polymer/ZnO hybrid solar cell — preparation and operation in air. *Nanotechnology*, **19(42)**, 424013, 12 p. (2008).

13. B. Winther-Jensen, M. Forsyth, K. West, J. W. Andreasen, P. Bayley, S. Pas, D. R. MacFarlane, Order–disorder transitions in poly(3,4-ethylenedioxythiophene). *Polymer*, **49(2)**, 481–487 (2008).

14. S. Goffri, C. Müller, N. Stingelin-Stutzmann, D. W. Breiby, C. P. Radano, J. W. Andreasen, R. Thompson, R. A. J. Janssen, M. M. Nielsen, P. Smith, H. Sirringhaus, Multicomponent semiconducting polymer systems with low crystallization-induced percolation threshold. *Nat. Mater.*, **5(12)**, 950–6 (2006).

15. J. W. Andreasen, S. A. Gevorgyan, C. M. Schlepütz, F. C. Krebs, Applicability of X-ray reflectometry to studies of polymer solar cell degradation. *Sol. Energy Mater. Sol. Cells*, **92(7)**, 793–798 (2008).

16. B. Paci, A. Generosi, V. R. Albertini, R. Generosi, P. Perfetti, R. de Bettignies, C. Sentein, Time-resolved morphological study of bulk heterojunction films for efficient organic solar devices. *J. Phys. Chem. C*, **112(26)**, 9931–9936 (2008).

17. P. Fenter, J. G. Catalano, C. Park, Z. Zhang, On the use of CCD area detectors for high-resolution specular X-ray reflectivity. *J. Synchrotron Radiat.*, **13(Pt 4)**, 293–303 (2006).

18. E. Luken, E. Ziegler, P. Høghøj, A. Freund, E. Gerdau, A. Fontaine, Growth Monitoring of W/Si X-Ray Multilayers by X-Ray Reflectivity and Kinetic Ellipsometry. *Opt. Interf. Coatings, Pts 1 2*, **2253**, 327–332 (1994).

19. D. W. Breiby, O. Bunk, J. W. Andreasen, H. T. Lemke, M. M. Nielsen, Simulating X-ray diffraction of textured films. *J. Appl. Crystallogr.*, **41(Part 2)**, 262–271 (2008).

20. R. Li, H. U. Khan, M. M. Payne, D.-M. Smilgies, J. E. Anthony, A. Amassian, Heterogeneous nucleation promotes carrier transport in solution-processed organic field-effect transistors. *Adv. Funct. Mater.*, **23(3)**, 291–297 (2013).

21. K. W. Chou, B. Yan, R. Li, E. Q. Li, K. Zhao, D. H. Anjum, S. Alvarez, R. Gassaway, A. Biocca, S. T. Thoroddsen, A. Hexemer, A. Amassian, Spin-cast bulk heterojunction solar cells: A dynamical investigation. *Adv. Mater.*, **25(13)**, 1923–9 (2013).

22. F. C. Krebs, Polymer solar cell modules prepared using roll-to-roll methods: Knife-over-edge coating, slot-die coating and screen printing. *Sol. Energy Mater. Sol. Cells*, **93(4)**, 465–475 (2009).

23. M. Sanyal, B. Schmidt-Hansberg, M. F. G. Klein, A. Colsmann, C. Munuera, A. Vorobiev, U. Lemmer, W. Schabel, H. Dosch, E. Barrena, In situ X-ray study of drying-temperature influence on the structural evolution of bulk-heterojunction polymer-fullerene solar cells processed by doctor-blading. *Adv. Energy Mater.*, **1(3)**, 363–367 (2011).

24. M. Sanyal, B. Schmidt-Hansberg, M. F. G. Klein, C. Munuera, A. Vorobiev, A. Colsmann, P. Scharfer, U. Lemmer, W. Schabel, H. Dosch, E. Barrena, Effect of photovoltaic polymer/fullerene blend composition ratio on microstructure evolution during film solidification investigated in real time by X-ray diffraction. *Macromolecules*, **44(10)**, 3795–3800 (2011).

25. B. Schmidt-Hansberg, M. Sanyal, M. F. G. Klein, M. Pfaff, N. Schnabel, S. Jaiser, A. Vorobiev, E. Müller, A. Colsmann, P. Scharfer, D. Gerthsen, U. Lemmer, E. Barrena, W. Schabel, Moving through the phase diagram: morphology formation in solution cast polymer-fullerene blend films for organic solar cells. *ACS Nano*, **5(11)**, 8579–90 (2011).

26. B. Schmidt-Hansberg, M. F. G. Klein, M. Sanyal, F. Buss, G. Q. G. de Medeiros, C. Munuera, A. Vorobiev, A. Colsmann, P. Scharfer, U. Lemmer, E. Barrena, W. Schabel,

Structure formation in low-bandgap polymer: Fullerene solar cell blends in the course of solvent evaporation. *Macromolecules*, **45(19)**, 7948–7955 (2012).

27. F. Liu, S. Ferdous, E. Schaible, A. Hexemer, M. Church, X. Ding, C. Wang, T. P. Russell, Fast printing and in situ morphology observation of organic photovoltaics using slot-die coating. *Adv. Mater.*, **27(5)**, 886–891 (2015).

28. A. P. L. Böttiger, M. Jørgensen, A. Menzel, F. C. Krebs, J. W. Andreasen, High-throughput roll-to-roll X-ray characterization of polymer solar cell active layers. *J. Mater. Chem.*, **22(42)**, 22501–22509 (2012).

29. J. Alstrup, M. Jørgensen, A. J. Medford, F. C. Krebs, Ultra fast and parsimonious materials screening for polymer solar cells using differentially pumped slot-die coating. *ACS Appl. Mater. Interfaces*, **2(10)**, 2819–27 (2010).

30. J. Kehres, J. W. Andreasen, F. C. Krebs, A. M. Molenbroek, I. Chorkendorff, T. Vegge, Combined in situ small- and wide-angle X-ray scattering studies of TiO(2) nanoparticle annealing to 1023 K. *J. Appl. Crystallogr.*, **43(Part 6)**, 1400–1408 (2010).

31. L. H. Rossander, N. K. Zawacka, H. F. Dam, F. C. Krebs, J. W. Andreasen, In situ monitoring of structure formation in the active layer of polymer solar cells during roll-to-roll coating. *AIP Adv.*, **4(8)**, 087105, 8 p. (2014).

32. N. K. Zawacka, T. R. Andersen, J. W. Andreasen, L. H. Rossander, H. F. Dam, M. Jørgensen, F. C. Krebs, The influence of additives on the morphology and stability of roll-to-roll processed polymer solar cells studied through ex situ and in situ X-ray scattering. *J. Mater. Chem. A*, **2(43)**, 18644–18654 (2014).

33. T. R. Andersen, H. F. Dam, M. Hösel, M. Helgesen, J. E. Carlé, T. T. Larsen-Olsen, S. A. Gevorgyan, J. W. Andreasen, J. Adams, N. Li, F. Machui, G. D. Spyropoulos, T. Ameri, N. Lemaître, M. Legros, A. Scheel, D. Gaiser, K. Kreul, S. Berny, O. R. Lozman, S. Nordman, M. Välimäki, M. Vilkman, R. R. Søndergaard, M. Jørgensen, C. J. Brabec, F. C. Krebs, Scalable, ambient atmosphere roll-to-roll manufacture of encapsulated large area, flexible organic tandem solar cell modules. *Energy Environ. Sci.*, **7(9)**, 2925–2933 (2014).

34. H. F. Dam, T. R. Andersen, E. B. L. Pedersen, K. T. S. Thydén, M. Helgesen, J. E. Carlé, P. S. Jørgensen, J. Reinhardt, R. R. Søndergaard, M. Jørgensen, E. Bundgaard, F. C. Krebs, J. W. Andreasen, Enabling flexible polymer tandem solar cells by 3D ptychographic imaging. *Adv. Energy Mater.*, **5(1)**, 1400736 (2015).

35. X. Gu, J. Reinspach, B. J. Worfolk, Y. Diao, Y. Zhou, H. Yan, K. Gu, S. Mannsfeld, M. F. Toney, Z. Bao, Compact roll-to-roll coater for in situ X-ray diffraction characterization of organic electronics printing. *ACS Appl. Mater. Interfaces*, **8(3)**, 1687–94 (2016).

36. L. H. Rossander, H. F. Dam, J. E. Carlé, M. Helgesen, I. Rajkovic, F. C. Krebs, J. W. Andreasen, In-line, roll-to-roll morphology analysis of organic solar cell active layers. *Energy Environ. Sci.*, **10**, 2411–2419 (2017).

37. M. A. Ruderer, E. Metwalli, W. N. Wang, G. Kaune, S. V Roth, P. Muller-buschbaum, thin films of photoactive polymer blends. *Chemphyschem*, **10(4)**, 664–671 (2009).

38. C. R. McNeill, B. Watts, L. Thomsen, W. J. Belcher, N. C. Greenham, P. C. Dastoor, H. Ade, Evolution of laterally phase-separated polyfluorene blend morphology studied by X-ray spectromicroscopy. *Macromolecules*, **42(9)**, 3347–3352 (2009).

Chapter 7

High Pressure Structural Studies of Conjugated Molecules

Matti Knaapila,[*,§] *Mika Torkkeli,*[*] *Ullrich Scherf,*[†] *Suchismita Guha*[‡]

*Department of Physics, Technical University of Denmark, 2800 Kgs. Lyngby, Denmark

†Macromolecular chemistry Group (buwmacro), Bergische Universität Wuppertal, 42119 Wuppertal, Germany

‡Department of Physics and Astronomy, University of Missouri, Columbia, MO 65211, USA

§matti.knaapila@fysik.dtu.dk

This chapter highlights high pressure GPa level structural studies of conjugated polymers and their analogues: conjugated oligomers and molecules, and rigid rod polymers. Attention is placed on our recent studies of polyfluorenes.

1. Introduction

This chapter discusses conjugated molecules under compression, which is mainly hydrostatic, following the ideas of Ref. 1. Conjugated molecules and oligomers and conjugated polymers are similar from intramolecular point of view. The former are monodisperse and usually highly crystalline with well-defined phase transitions. The latter are polydisperse with a limited persistence length and smooth transitions. Rigid rod polymers and conjugated polymers are similar from intermolecular point of view and manifest for example lyotropic and thermotropic liquid crystalline behaviour. We first discuss the structural properties of conjugated molecules and rigid rod polymers under compression. Then, we discuss conjugated polymers under compression beginning with polythiophenes and ending up with our recent studies of poly[9,9-bis(2-ethylhexyl)fluorene] (PF2/6). Finally, we conclude some essential similarities and differences between these materials.

2.   Conjugated Molecules

The pressure induced structural effects have been reported for several aromatic or heterocyclic molecules including indole,[2] diphenyl-1,2,3-oxadiazoles,[3] or bis-benzene-1,2-dithiolato crystals,[4] oligoacenes including anthracene,[5] pentacene, and tetracene[6] as well as for conjugated oligomers such as oligo(*p*-phenylene)s including biphenyl[7] and terphenyl.[8]

These studies have identified two factors that are generically important for compressed conjugated molecules: First, the molecules are anisotropic both in structure and compression. Second, they may become planarized under compression if the molecules allow internal rotation between subsequent moieties like in the case of oligo *p*-phenylenes. The planarization is characterized by the torsion angle τ (or sometimes marked as φ) between planar moieties, and eventually affects the planarity of the whole molecule. Likewise, anisotropy and planarization are essential for compressed conjugated polymers

Conjugated molecules manifest both gradual compressions of existing phases and also transitions to completely new high pressure phases. For example, ambient fluorene shows an orthorhombic unit cell consisting of four herringbone positioned molecules[9] and this phase transforms to the π-stacked high pressure phase with compression.[10] Figure 1 plots X-ray diffraction (XRD) curves of compressed fluorene[10] showing the phase transition around 3.6 GPa.

3.   Rigid Rod Polymers and Conjugated Rigid Rod Polymers

Polyimides are archetypical rigid rod polymers. In one particular example, the composition of polyimides was varied in terms of phenyl rings in the diamine moiety (and thus effectively in terms of the rigidity of the system).[11–14] When these materials were compressed, the compressibility along the main chain (c-axis) was found to increase with the increasing number of phenyl rings. Moreover, when the number of phenyl rings was increased the compression turned from isotropic to anisotropic along interchain directions (a- and b-axes). These results are illustrated in Figure 2.

Poly(*p*-phenylenes) (PPPs) represent archetypical conjugated rigid rod polymers, and with alkyl side chains they are recognized as hairy-rod polymers.

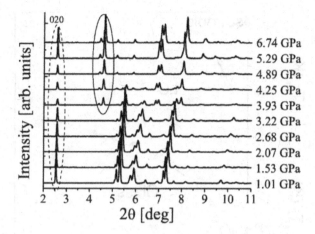

Figure 1: XRD curves of fluorene crystal with increasing pressure. Dashed and solid ellipses illustrate 020 reflection associated with the fluorene layers and a new phase between 3.2 GPa and 3.9 GPa. (Reproduced with permission from Ref. 10. Copyright 2006 The American Physical Society.)

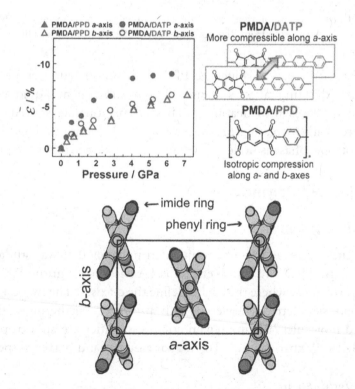

Figure 2: Strain along the crystallographic a- and b-axes for two different polyimides PMDA/DATP and PMDA/PPD with different number of phenyl moieties. The a- and b-axes are parallel and perpendicular to the stacking direction and both perpendicular to the rigid polymer chain that coincidences with the c-axis. (Reproduced with permission from Ref. 14. Copyright 2014 The American Chemical Society.)

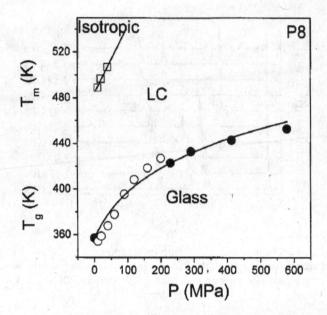

Figure 3: Phase diagram of rigid rod PPP as a function of temperature and pressure. (Reproduced with permission from Ref. 15. Copyright 2004 The American Physical Society.)

Characteristic for hairy-rods include PPPs with sulfonated ester and dodecyl side chains exhibiting a glass transition and subsequent liquid crystalline and isotropic phases with heating.[15] These transitions are shifted to higher temperatures with increasing pressure as shown in Figure 3. This also shows that the isotropic phase is no longer attainable for sufficiently high pressures.

4.   Conjugated Polymers

4.1.   *Polyacetylene*

Polyacetylene (PA) was the first conjugated polymer that was studied under compression by XRD.[16–18] Ambient trans-PA has a space group P21/n and a monoclinic structure where the chains define the c-axis.[19] The two-dimensional space group symmetry is lowered with pressure. Furthermore, like with conjugated molecules, the interchain compressibilities are anisotropic having values $8.3 \cdot 10^{-3}$/kbar and $3.8 \cdot 10^{-3}$/kbar along a- and b-axes, respectively.[16]

4.2.   *Polythiophene*

These early examples were followed by alkyl-substituted polythiophenes with well-known ambient structures.[21] Poly(3-octylthiophene) (POT), for example,

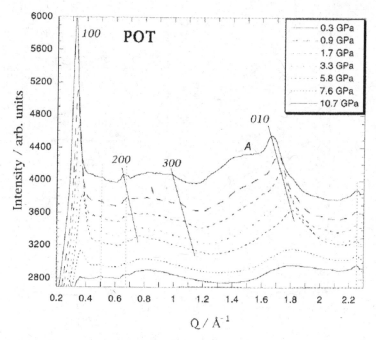

Figure 4: XRD curves of POT with *h*00 and 010 reflections with increasing pressure. (Reproduced from Ref. 20. Copyright 1998 The Institute of Physics.)

manifests orthorhombic lamellae of backbones and side chains. In this structure a-, b- and c-axes are defined by the layer normal, backbone stack, and main chain. The XRD curves show distinctive *h*00 peaks and broader 010 (sometimes indexed as 020) peak on top of the amorphous feature plus weak 001 at a wider angle.[22] Figure 4 shows XRD data of POT with increasing pressure.[20] Both *h*00 and 010 reflections move towards higher scattering angles which points to the decreasing lamellar period and increasing backbone planarity. Decreasing distance between neighbouring chains increases their planarity and this leads to an optical transition from red to yellow.[20,23–25]

Lately much attention was placed on polythiophenes blended with methanofullerenes and particularly on poly(3-hexylthiophene) (P3HT) blended with phenyl-C61-butyric acid methyl ester (PCBM). They are mixed in the molecular level but remain segregated into polymer and fullerene rich microphases where crystallinity depends on the chose lend ratio.[27] When compressed, this blend shows a decreased LUMO and badgap which are attributed to the polymer planarization.[28]

Figure 5 shows XRD curves of P3HT:PCBM blend with increasing pressure indicating decreasing crystallite size and confirming planarization of P3HT.[26]

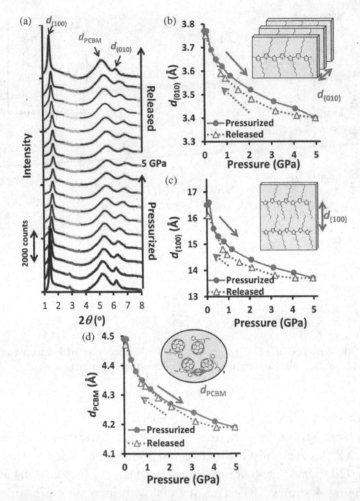

Figure 5:   (a) XRD curves of P3HT:PCBM (1:1) blend as a function of pressure. The data show 100 and 010 reflections for P3HT and the reflection associated with the intermolecular distance for PCBM. (b–d) The stacking and interlamellar distances for compressed P3HT and intermolecular distance for compressed PCBM. (Reproduced with permission from Ref. 26. Copyright 2015 The American Chemical Society.)

The compression is anisotropic likening rigid rod polymer data shown in Figure 2.

4.3.   *Polyfluorene*

We have recently placed attention on the high pressure studies of helical PF2/6[29–31] as a structural isomer of poly(9,9-dioctylfluorene) PFO.[32] The intermolecular structure of PF2/6 depends on the molecular weight.

A limiting molecular weight M^* differentiates nematic low molecular weight PF2/6 (LMW-PF2/6) and hexagonal high molecular weight PF2/6 (HMW-PF2/6).[33] Similarly, hexagonal HMW-PF2/6 becomes nematic with heating. The LMW nematic phase consists of polymer 3-mers and HMW nematic phase of individual polymers. Ambient LMW nematic is the densest form followed by hexagonal and nematic HMW phases. Compression has two kinds of effects on PF2/6: Intermolecular and intramolecular.

The ambient intermolecular hexagonal order is impaired at or below 2 GPa resembling an ambient transition from hexagonal to nematic phase[34] concomitant with significant optical changes.[35] However, the ambient transition occurs above glass transition whereas pressure induced loss of long range order occurs below the glass transition (and we denote the obtained structure as glassy nematic phase). Furthermore, the pressure induced transition must be towards denser phase possibly incorporating polymer 3-mers like the ambient LMW nematic phase. In contrast, already nematic LMW-PF2/6 maintains its comparably weaker long range order under compression.[36]

The ambient intramolecular order is characterized by XRD and 00*l* reflections that describe the polymer helicity and the unit cell along c-axis. Both HMW and LMW PF2/6 show a slight deviation in the 00*l* reflections towards higher scattering angles with increasing pressure.

To consider PF2/6 planarization under pressure we have built a model that includes *trans* or "near"-*trans* torsion angles and allow at least 4-fold and 2-fold periodicities.[37,38] These considerations are illustrated in Figure 6. The first two models are 4/1 and 2/1 helixes and the third geometry $(TS)_2$ has alternating *trans* or "near"-*trans* torsion angles. The fourth geometry $(GS)_2$ has two rotations where the third monomer is in the *trans* position compared to the first. The two rotations satisfy the condition $\tau_s + \tau_g \sim 180°$. The two last geometries are twisted conformations through alternating left-hand right-hand rotations τs and $-\tau$s. Characteristically, all these models manifest symmetries about the c-axis and produce the 00*l* reflections whose positions describe the torsion angle.

Figure 7 plots the peak positions for helical structures and for the $(GS)_2$ and $(TS)_2$ geometries.[37] Any pressure induced planarization should decrease torsion angle τ and seen as a shift of 00l reflections towards higher scattering angles.

Figure 8 shows XRD data of PF2/6 under compression and decompression. The data are dominated by a 00*l* which moves towards higher scattering angles with increasing pressure and returns with decreasing pressure. Figure 9 shows the 00*l* reflection fitted to these data and interpreted using a model built on

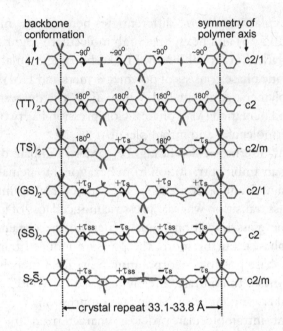

Figure 6: Models for the polyfluorene backbone conformations that produce the 00l reflections. The models manifest different symmetries about the c-axis and are characterized by torsion angles ~90°, 180°, $\pm\tau_s$, and $\pm\tau_g$. T = trans. S = skew. G = gauche. (Reproduced with permission from.[37] Copyright 2013 The American Physical Society.)

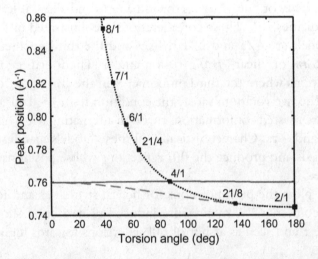

Figure 7: Calculated 00l reflection as a function of torsion angle for various helixes (dotted black line and square markers) and non-helical structures where base unit consist of two (solid blue line — (GS)$_2$) or four alternating torsion angles (dashed red line- (TS)$_2$). (Reproduced with permission from.[37] Copyright 2013 The American Physical Society.)

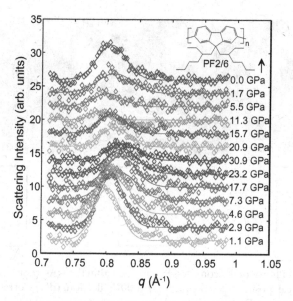

Figure 8: XRD curves (diamonds) and fits to the data (solid lines) of PF2/6 with increasing and decreasing pressure. (Reproduced with permission from Ref. 38. Copyright 2013 The American Chemical Society.)

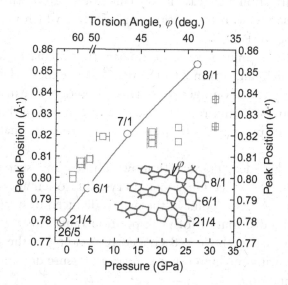

Figure 9: The most prominent 00*l* reflection as a function of pressure corresponding to the data shown in Figure 8 (red squares). The peak position for selected theoretical PF2/6 helixes (open circles) as a function of torsion angle, φ. (Reproduced with permission from Ref. 38. Copyright 2013 The American Chemical Society.)

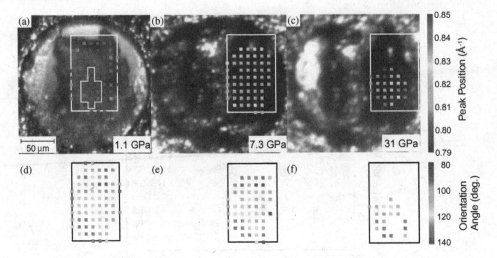

Figure 10:   (a–c) Photos of the compressed samples within the DAC gasket at various pressures and maps of the peak positions for the prominent 00*l* reflection. (d–f) Corresponding maps for the preferred orientation direction. (Reproduced with permission from Ref. 38. Copyright 2013 The American Chemical Society.)

helices with decreasing τ. These data indicate that the pressure leads to the backbone planarization (decrease in τ). This result agrees with the minimum energy conformation with the torsion angle at 43°, which is calculated from the first principles in Ref. 30.

The hydrostatic limit of pressure transmitting media is defined by their solidification which range from 7 to 15 GPa.[39] Above these values the pressure distribution becomes inhomogeneous. This inhomogeneity may be treated by mapping samples with an x-ray microbeam and comparing the obtained data distribution to the pressure distribution obtained from the analysis of pressure standards.

Following this idea, Figure 10 shows photos and spatial maps of the most prominent 00*l* reflection and the crystallite orientation for PF2/6 at selected pressures.[38] Each point in the map is deduced from a dataset plotted in Figure 8. The distribution of peak position (and thus the torsion angle) is significantly widened above 7–8 GPa corresponding to the solidification of pressure transmitting medium (neon). Yet the pressure does not influence the crystallite orientation.

5.   Conclusions

Conjugated molecules, oligomers and polymers as well as conjugated rigid rod polymers share structural phenomena and viewpoints under extreme pressure conditions research. These include structural anisotropy and anisotropic effect of pressure, molecular planarization and impairment of long range order as well as shift of phase transitions. Much of the structural characteristics of conjugated polymers stem from their stiff backbone and therefore when pursuing high pressure research, it is equally illustrative to make comparisons across rigid rod polymers and hairy-rod polymers whether conjugated or not.

Future research in this area should consolidate the different structural aspects across a range of conjugated molecules and polymers, as discussed in this chapter, and also characterize structures and phase behaviour across MPa and GPa pressure ranges. Furthermore, attention should be placed on macroscopic anisotropy that may be followed both by X-rays and by polarized optical emission as shown in a rare example for compressed poly(*p*-phenylene vinylene).[40]

References

1. M. Knaapila, S. Guha, Blue emitting organic semiconductors under high pressure: Status and outlook. *Rep. Prog. Phys.*, **79**, 066601 (2016).
2. B. Schatschneider, J. J. Lian, Simulated pressure response of crystalline indole. *J. Chem. Phys.*, **135**, 164508 (2011).
3. I. Orgzall, F. Emmerling, B. Schulz, O. Franco, High-pressure studies on molecular crystals — relations between structure and high-pressure behavior. *J. Phys.: Condes. Matter*, **20**, 295206 (2008).
4. N. C. Schiødt, Y. Bjørnholm, K. Bechgaard, J. J. Neumeier; C. Allgeier, C. S. Jacobsen, N. Thorup, Structural, electrical, magnetic, and optical properties of bis-benzene-1,2-dithiolato-au(iv) crystals. *Phys. Rev. B*, **53**, 1773 (1996).
5. M. Oehzelt, G. Heimel, R. Resel, P. Puschnig, K. Hummer, C. Ambrosch-Draxl, K. Takemura, A. Nakayama, High pressure x-ray study on anthracene. *J. Chem. Phys.*, **119**, 1078 (2003).
6. M. Oehzelt, A. Aichholzer, R. Resel, G. Heimel, E. Venuti, R. G. Della Valle, Crystal structure of oligoacenes under high pressure. *Phys. Rev. B*, **74**, 104103 (2006).
7. P. Puschnig, C. Ambrosch-Draxl, G. Heimel, E. Zojer, R. Resel, G. Leising, M. Kriechbaum, W. Graupner, Pressure studies on the intermolecular interaction in biphenyl. *Synth. Met.*, **116**, 327 (2001).
8. P. Puschnig, K. Hummer, C. Ambrosch-Draxl, G. Heimel, M. Oehzelt, R. Resel, Electronic, optical, and structural properties of oligophenylene molecular crystals under high pressure: An *ab initio* investigation. *Phys. Rev. B*, **67**, 235321 (2003).
9. V. K. Belsky, V. E. Zavodnik, V. M. Vozzhenikov, Fluorene, c13h10. *Acta Cryst. C*, **40**, 1210 (1984).
10. G. Heimel, K. Hummer, C. Ambrosch-Draxl, W. Chunwachirasiri, M. J. Winokur, M. Hanfland, M. Oehzelt, A. Aichholzer, R. Resel, Phase transition and electronic properties of fluorene: A joint experimental and theoretical high-pressure study. *Phys. Rev. B*, **73**, 024109 (2006).

11. K. Takizawa, J. Wakita, M. Kakiage, H. Masunaga, S. Ando, Molecular aggregation of polyimide films at very high pressure analyzed by synchrotron wide-angle x-ray diffraction. *Macromolecules*, **43**, 2115 (2010).

12. K. Takizawa, J. Wakita, S. Azami, S. Ando, Relationship between molecular aggregation structures and optical properties of polyimide films analyzed by synchrotron wide-angle x-ray diffraction, infrared absorption, and uv/visible absorption spectroscopy at very high pressure. *Macromolecules*, **44**, 349 (2011).

13. K. Takizawa, J. Wakita, K. Sekiguchi, S. Ando, Variations in aggregation structures and fluorescence properties of a semialiphatic fluorinated polyimide induced by very high pressures. *Macromolecules*, **45**, 4764 (2012).

14. K. Takizawa, H. Fukudome, Y. Kozaki, S. Ando, Pressure-induced changes in crystalline structures of polyimides analyzed by wide-angle x-ray diffraction at high pressures. *Macromolecules*, **47**, 3951 (2014).

15. A. Gitsas, G. Floudas, G. Wegner, Effects of temperature and pressure on the stability and mobility of phases in rigid rod poly(p-phenylenes). *Phys. Rev. E*, **69**, 041802 (2004).

16. J. Ma, J. E. Fischer, Y. Cao, A. J. Heeger, X-ray structural study of *trans*-polyacetylene at high pressure. *Solid State Comm.*, **83**, 395 (1992).

17. P. Papanek, J. E. Fischer, Molecular-dynamics simulation of crystalline *trans*-polyacetylene. *Phys. Rev. B*, **48**, 12566 (1993).

18. A. Matsushita, K. Akagi, T.-S. Liang, H. Shirakawa, Effects of pressure on the electrical resistivity of iodine doped polyacetylene. *Synth. Met.*, **101**, 447 (1999).

19. Q. Zhu, J. E. Fischer, Crystal structure of polyacetylene revisited: An x-ray study. *Solid State Comm.*, **83**, 179 (1992).

20. J. Mårdalen, E. J. Samuelsen, O. R. Konestabo, M. Hanfland, M. Lorenzen, Conducting polymers under pressure: Synchrotron x-ray determined structure and structure related properties of two forms of poly(octyl-thiophene). *J. Phys.: Condens. Matter*, **10**, 7145 (1998).

21. M. J. Winokur, Structural studies of conducting polymers. In *Handbook of Conducting polymers*; T. A. Skotheim, R. L. Elsenbaumer, J. R. Reynolds, Eds.; Marcel Dekker: New York, 1998; pp 707.

22. T. J. Prosa, M. J. Winokur, J. Moulton, P. Smith, A. J. Heeger, X-ray-diffraction studies of the three-dimensional structure with iodine-intercalated poly(3-octylthiophene). *Phys. Rev. B*, **51**, 159 (1995).

23. J. Mårdalen, Y. Cerenius, P. Häggkvist, The crystalline structure of poly(3-octylthiophene) at high pressure. *J. Phys.: Condens. Matter*, **7**, 3501 (1995).

24. J. Corish, D. A. Morton-Blake, F. Bénière, M. Lantoine, Interaction of side-chains in poly(3-alkylthiophene) lattices. *J. Chem. Soc., Faraday Trans.*, **92**, 671 (1996).

25. S. O'Dwyer, H. Xie, J. Corish, D. A. Morton-Blake, An atomistic simulation of the effect of pressure on conductive polymers. *J. Phys.: Condens. Matter*, **13**, 2395 (2001).

26. Y. Noguchi, A. Saeki, T. Fujiwara, S. Yamanaka, M. Kumano, T. Sakurai, N. Matsuyama, M. Nakano, N. Hirao, Y. Ohishi, S. Seki, Pressure modulation of backbone conformation and intermolecular distance of conjugated polymers toward understanding the dynamism of π-figuration of their conjugated system. *J. Phys. Chem. B*, **119**, 7219 (2015).

27. T. Agostinelli, S. Lilliu, J. G. Labram, M. Campoy-Quilles, M. Hampton, E. Pires, J. Rawle, O. Bikondoa, D. D. C. Bradley, T. D. Anthopoulos, J. Nelson, J. E. Macdonald, Real-time

investigation of crystallization and phase-segregation dynamics in p3ht:Pcbm solar cells during thermal annealing. *Adv. Funct. Mater.*, **21**, 1701 (2011).

28. K. Paudel, M. Chandrasekhar, U. Scherf, E. Preis, S. Guha, High-pressure optical studies of donor-acceptor polymer heterojunctions. *Phys. Rev. B*, **84**, 205208 (2011).

29. G. Lieser, M. Oda, T. Miteva, A. Meisel, H.-G. Nothofer, U. Scherf, D. Neher, Ordering, graphoepitaxial orientation, and conformation of a polyfluorene derivative of the "Hairy-rod" Type on an oriented substrate of polyimide. *Macromolecules*, **33**, 4490 (2000).

30. B. Tanto, S. Guha, C. M. Martin, U. Scherf, M. J. Winokur, Structural and spectroscopic investigations of bulk poly[bis(2-ethyl)hexylfluorene]. *Macromolecules*, **37**, 9438 (2004).

31. M. Knaapila, M. Torkkeli, A. P. Monkman, Evidence for 21-helicity of poly[9,9-bis(2-ethylhexyl)fluorene-2,7-diyl]. *Macromolecules*, **40**, 3610 (2007).

32. S. McFarlane, R. McDonald, J. G. C. Veinot, 9,9-di-n-octyl-9h-fluorene. *Acta Cryst. C*, **61**, o671 (2005).

33. M. Knaapila, R. Stepanyan, M. Torkkeli, B. P. Lyons, T. P. Ikonen, L. Almásy, J. P. Foreman, R. Serimaa, R. Güntner, U. Scherf, A. P. Monkman, Influence of molecular weight on the phase behavior and structure formation of branched side chain hairy-rod polyfluorene in bulk phase. *Phys. Rev. E*, **71**, 041802 (2005).

34. M. Knaapila, R. Stepanyan, D. Haase, S. Carlson, M. Torkkeli, Y. Cerenius, U. Scherf, S. Guha, Evidence for structural transition in hairy-rod poly[9,9-bis(2-ethylhexyl)fluorene] under high pressure conditions. *Phys. Rev. E*, **82**, 051803 (2010).

35. C. M. Martin, S. Guha, M. Chandrasekhar, H. R. Chandrasekhar, R. Guentner, P. Scanduicci de Freitas, U. Scherf, Hydrostatic pressured dependence of the luminescence and raman frequencies in polyfluorene. *Phys. Rev. B*, **68**, 115203 (2003).

36. S. Guha, M. Knaapila, D. Moghe, Z. Konôpková, M. Torkkeli, M. Fritsch, U. Scherf, Persistence of nematic liquid crystalline phase in a polyfluorene-based organic semiconductor: A high pressure study. *J. Polym. Sci. Part B Polym. Phys.*, **52**, 1014 (2014).

37. M. Knaapila, Z. Konôpková, M. Torkkeli, D. Haase, H.-P. Liermann, S. Guha, U. Scherf, Structural study of helical polyfluorene under high quasihydrostatic pressure. *Phys. Rev. E*, **87**, 022602 (2013).

38. M. Knaapila, M. Torkkeli, Z. Konôpková, D. Haase, H.-P. Liermann, U. Scherf, S. Guha, Measuring structural inhomogeneity of conjugated polymer at high pressures up to 30 gpa. *Macromolecules*, **46**, 8284 (2013).

39. S. Klotz, J.-C. Chervin, P. Munsch, G. Le Marchand, Hydrostatic limits of 11 pressure transmitting media. *J. Phys. D: Appl. Phys.*, **42**, 075413 (2009).

40. V. Morandi, M. Galli, F. Marabelli, D. Comoretto, Highly oriented poly(paraphenylene vinylene): Polarized optical spectroscopy under pressure. *Phys. Rev. B*, **79**, 045202 (2009).

Index

World Scientific Series in Materials and Energy

(Continuation of series card page)

Printed in the United States
By Bookmasters